DEFENSE, SECURITY AND STRATEGIES

DOMESTIC DRONES

ELEMENTS AND CONSIDERATIONS FOR THE U. S.

DEFENSE, SECURITY AND STRATEGIES

Additional books in this series can be found on Nova's website under the Series tab.

Additional e-books in this series can be found on Nova's website under the e-book tab.

DEFENSE, SECURITY AND STRATEGIES

DOMESTIC DRONES

ELEMENTS AND CONSIDERATIONS FOR THE U. S.

DONOVAN AMBROSIO
EDITOR

nova publishers
New York

For permission to use material from this book please contact us:
Telephone 631-231-7269; Fax 631-231-8175
Web Site: http://www.novapublishers.com

NOTICE TO THE READER

Library of Congress Cataloging-in-Publication Data

ISBN: 978-1-63321-456-9

Published by Nova Science Publishers, Inc. † New York

CONTENTS

PREFACE

The integration of drones into U. S. skies is expected by many to yield significant commercial and societal benefits. Drones could be employed to inspect pipelines, survey crops, and monitor the weather. One newspaper has already used a drone to survey storm damage, and real estate agents have used them to survey property. In short, the extent of their potential domestic application is bound only by human ingenuity. This book explores some of the legal issues that will inevitably generate with the introduction of drones into U. S. skies. This book also discusses the fourth amendment's implications and legislative responses; background and considerations for Congress of drones in domestic surveillance operations and pilotless drones; and provides a report to Congress on future unmanned aircraft systems training, operations, and sustainability.

Chapter 1 - Under the FAA Modernization and Reform Act of 2012, P.L. 112-95, Congress has tasked the Federal Aviation Administration (FAA) with integrating unmanned aircraft systems (UASs), sometimes referred to as unmanned aerial vehicles (UAVs) or drones, into the national airspace system by September 2015. Although the text of this act places safety as a predominant concern, it fails to address significant, and up to this point, largely unanswered legal questions.

For instance, several legal interests are implicated by drone flight over or near private property. Might such a flight constitute a trespass? A nuisance? If conducted by the government, a constitutional taking? In the past, the Latin maxim *cujus est solum ejus est usque ad coelum* (for whoever owns the soil owns to the heavens) was sufficient to resolve many of these types of questions, but the proliferation of air flight in the 20th century has made this proposition untenable. Instead, modern jurisprudence concerning air travel is significantly more nuanced, and often more confusing. Some courts have relied on the federal definition of "navigable airspace" to determine which flights could constitute a trespass. Others employ a nuisance theory to ask whether an overhead flight causes a substantial impairment of the use and enjoyment of one's property. Additionally, courts have struggled to determine when a government-operated overhead flight constitutes a taking under the Fifth and Fourteenth Amendments.

With the ability to house surveillance sensors such as high-powered cameras and thermal-imaging devices, some argue that drone surveillance poses a significant threat to the privacy of American citizens. Because the Fourth Amendment's prohibition against unreasonable searches and seizures applies only to acts by government officials, surveillance by private actors such as the paparazzi, a commercial enterprise, or one's neighbor is instead regulated, if at all, by state and federal statutes and judicial decisions. Yet, however strong this interest in privacy may be, there are instances where the public's First Amendment rights to *gather* and *receive* news might outweigh an individual's interest in being let alone.

Additionally, there are a host of related legal issues that may arise with this introduction of drones in U. S. skies. These include whether a property owner may protect his property from a trespassing drone; how stalking, harassment, and other criminal laws should be applied to acts committed with the use of drones; and to what extent federal aviation law could preempt future state law.

Because drone use will occur largely in federal airspace, Congress has the authority or can permit various federal agencies to set federal policy on drone use in American skies. This may include the appropriate level of individual privacy protection, the balancing of property interests with the economic needs of private entities, and the appropriate safety standards required.

Chapter 2 - The prospect of drone use inside the United States raises far-reaching issues concerning the extent of government surveillance authority, the value of privacy in the digital age, and the role of Congress in reconciling these issues.

Drones, or unmanned aerial vehicles (UAVs), are aircraft that can fly without an onboard human operator. An unmanned aircraft system (UAS) is the entire system, including the aircraft, digital network, and personnel on the ground. Drones can fly either by remote control or on a predetermined flight path; can be as small as an insect and as large as a traditional jet; can be produced more cheaply than traditional aircraft; and can keep operators out of harm's way. These unmanned aircraft are most commonly known for their operations overseas in tracking down and killing suspected members of Al Qaeda and related organizations. In addition to these missions abroad, drones are being considered for use in domestic surveillance operations to protect the homeland, assist in crime fighting, disaster relief, immigration control, and environmental monitoring.

Although relatively few drones are currently flown over U. S. soil, the Federal Aviation Administration (FAA) predicts that 30,000 drones will fill the nation's skies in less than 20 years. Congress has played a large role in this expansion. In February 2012, Congress enacted the FAA Modernization and Reform Act (P.L. 112-95), which calls for the FAA to accelerate the integration of unmanned aircraft into the national airspace system by 2015. However, some Members of Congress and the public fear there are insufficient safeguards in place to ensure that drones are not used to spy on American citizens and unduly infringe upon their fundamental privacy. These observers caution that the FAA is primarily charged with ensuring air traffic safety, and is not adequately prepared to handle the issues of privacy and civil liberties raised by drone use.

This report assesses the use of drones under the Fourth Amendment right to be free from unreasonable searches and seizures. The touchstone of the Fourth Amendment is reasonableness. A reviewing court's determination of the reasonableness of a drone search would likely be informed by location of the search, the sophistication of the technology used, and society's conception of privacy in an age of rapid technological advancement. While individuals can expect substantial protections against warrantless government intrusions into their homes, the Fourth Amendment offers less robust restrictions upon government surveillance occurring in public places including areas immediately outside the home, such as in driveways or backyards. Concomitantly, as technology advances, the contours of what is reasonable under the Fourth Amendment may adjust as people's expectations of privacy evolve.

Chapter 3 - Growing interest in the use of unmanned aerial vehicles (UAVs), particularly for homeland security and law enforcement applications, has spurred considerable debate over how to accommodate these unmanned aircraft and keep them safely separated from other air traffic. Additionally, the use of these pilotless aircraft, popularly referred to as drones, for aerial surveillance and law enforcement purposes has raised specific concerns regarding privacy and Fourth Amendment rights and potential intrusiveness. These issues have come to the forefront in policy debate in response to provisions in the FAA Modernization and Reform Act of 2012 (P.L. 112-95) that require the Federal Aviation Administration (FAA) to begin integrating unmanned aircraft into the national airspace system by the end of FY2015.

While drones have been used extensively by the military and small radio-controlled model aircraft have been around for more than 50 years, advances in more complex vehicle controls and imaging sensor capabilities are spurring public sector and commercial interest in unmanned aircraft for a variety of purposes, including law enforcement, homeland security, aerial imaging, and scientific research. FAA currently approves public entities (such as federal agencies, public universities, and local police departments) to operate UAVs on a case-by-case basis, but growing interest is making this approach increasingly untenable. Moreover, commercial users are seeking authorization to fly drones, but so far FAA has only allowed test and demonstration flights by manufacturers. FAA faces a number of challenges to address anticipated growth in demand for civilian UAV operations and develop regulations governing the certification and operation of unmanned aircraft systems in domestic airspace.

Congress has generally supported efforts to integrate unmanned aircraft into the national airspace system and foster growth in the unmanned aircraft industry. It enacted extensive provisions in P.L. 112-95 that are designed to streamline and accelerate the operation of unmanned aircraft in domestic airspace by both public entities and commercial operators. Notably, that law requires FAA to issue regulations pertaining to the operation of small UAVs (weighing less than 55 pounds) and requires FAA to create and implement a plan to begin the integration of unmanned aircraft into the national airspace system by the end of FY2015. Toward that goal, the act requires FAA to establish six test ranges throughout the United States to study unmanned aircraft integration technical issues.

The act establishes an ambitious timeline for FAA to grapple with and resolve a number of complex issues regarding the safety and security of unmanned aircraft operations. Furthermore, aircraft operators have expressed specific concerns that drone operations should not result in airspace restrictions or other measures that could limit accessibility to the national airspace system.

In addition to these various challenges, the privacy implications and potential intrusiveness of drone operations have emerged as a significant issue before Congress. Civil liberties and privacy groups have cautioned that voluntary industry measures, including a code of conduct to, among other things, respect privacy, are inadequate to assure that drones will not be misused in ways that could infringe upon the privacy of individuals and intrude upon their daily activities. Moreover, FAA's authority over specific uses of civilian unmanned aircraft appears limited so long as safety and national security are not compromised, raising additional concerns that future drone operations could lead to complaints and lawsuits over noise, intrusiveness, and interference with the use and enjoyment of public or private property.

Chapter 4 - This report is being provided to the congressional defense committees as requested in House Armed Services Committee Report 111-491, accompanying the Ike Skelton National Defense Authorization Act for Fiscal Year 2011.

TASK: Report on Future Unmanned Aerial Systems Training, Operations, and Sustainability. The rapid growth of UAS inventories to meet operational demands raises a number of questions concerning the military services' ability to support these inventories in the near-and long-term. In particular, to support their UAS inventories, the military services must train sufficient numbers of personnel to operate and maintain the aircraft, provide adequate facilities and other infrastructure to sustain them, and provide sufficient access to airspace and training ranges to train military personnel within the United States and at military bases overseas.

The committee directs the Secretary of Defense to provide a report to the congressional defense committees with its fiscal year 2012 budget request that describes the military services' plans to support their current and planned UAS inventories. The report should, at a minimum, discuss:

1. Current UAS inventory levels and planned UAS inventory levels for each fiscal year through 2017;
2. Plans to supply the number of personnel needed to operate the aircraft and sensor payloads and to perform UAS maintenance;
3. Current and planned UAS basing and other operating locations;
4. Progress made in providing the number of facilities needed for UAS inventories to support operations and training and the funding required for any additional facilities; and
5. The availability of airspace, ranges, and other infrastructure at each planned UAS location, and a description of the steps that the services plan to take to overcome any limitations that adversely impact UAS training.

In: Domestic Drones
Editor: Donovan Ambrosio

ISBN: 978-1-63321-456-9

Chapter 1

INTEGRATION OF DRONES INTO DOMESTIC AIRSPACE: SELECTED LEGAL ISSUES*

Alissa M. Dolan and Richard M. Thompson II

SUMMARY

Under the FAA Modernization and Reform Act of 2012, P.L. 112-95, Congress has tasked the Federal Aviation Administration (FAA) with integrating unmanned aircraft systems (UASs), sometimes referred to as unmanned aerial vehicles (UAVs) or drones, into the national airspace system by September 2015. Although the text of this act places safety as a predominant concern, it fails to address significant, and up to this point, largely unanswered legal questions.

For instance, several legal interests are implicated by drone flight over or near private property. Might such a flight constitute a trespass? A nuisance? If conducted by the government, a constitutional taking? In the past, the Latin maxim *cujus est solum ejus est usque ad coelum* (for whoever owns the soil owns to the heavens) was sufficient to resolve many of these types of questions, but the proliferation of air flight in the 20th century has made this proposition untenable. Instead, modern jurisprudence concerning air travel is significantly more nuanced, and often more confusing. Some courts have relied on the federal definition of "navigable airspace" to determine which flights could constitute a trespass. Others employ a nuisance theory to ask whether an overhead

---

* This is an edited, reformatted and augmented version of a Congressional Research Service publication, No. R42940, dated April 4, 2013.

flight causes a substantial impairment of the use and enjoyment of one's property. Additionally, courts have struggled to determine when a government-operated overhead flight constitutes a taking under the Fifth and Fourteenth Amendments.

With the ability to house surveillance sensors such as high-powered cameras and thermal-imaging devices, some argue that drone surveillance poses a significant threat to the privacy of American citizens. Because the Fourth Amendment's prohibition against unreasonable searches and seizures applies only to acts by government officials, surveillance by private actors such as the paparazzi, a commercial enterprise, or one's neighbor is instead regulated, if at all, by state and federal statutes and judicial decisions. Yet, however strong this interest in privacy may be, there are instances where the public's First Amendment rights to *gather* and *receive* news might outweigh an individual's interest in being let alone.

Additionally, there are a host of related legal issues that may arise with this introduction of drones in U. S. skies. These include whether a property owner may protect his property from a trespassing drone; how stalking, harassment, and other criminal laws should be applied to acts committed with the use of drones; and to what extent federal aviation law could preempt future state law.

Because drone use will occur largely in federal airspace, Congress has the authority or can permit various federal agencies to set federal policy on drone use in American skies. This may include the appropriate level of individual privacy protection, the balancing of property interests with the economic needs of private entities, and the appropriate safety standards required.

INTRODUCTION

The integration of drones into U. S. skies is expected by many to yield significant commercial and societal benefits.[1] Drones could be employed to inspect pipelines, survey crops, and monitor the weather.[2] One newspaper has already used a drone to survey storm damage,[3] and real estate agents have used them to survey property.[4] In short, the extent of their potential domestic application is bound only by human ingenuity.

In an effort to accelerate this introduction, in the FAA Modernization and Reform Act of 2012, Congress tasked the Federal Aviation Administration (FAA) with safely integrating drones into the national airspace system by September 2015.[5] Likewise, sensing the opportunities that unmanned flight portend, lobbying groups and drone manufacturers have joined the chorus of those seeking a more rapid expansion of drones in the domestic market.[6]

Yet, the full-scale introduction of drones into U. S. skies will inevitably generate a host of legal issues. This report will explore some of those issues. To begin, this report will describe the regulatory framework for permitting the use of unmanned vehicles and the potential rulemaking that will occur over the next few years. Next, it will discuss theories of takings and property torts as they relate to drone flights over or near private property. It will then discuss the privacy interests implicated by drone surveillance conducted by private actors and the potential countervailing First Amendment rights to gather and receive news. Finally, this report will explore possible congressional responses to these privacy concerns, discuss how the FAA has approached these concerns, and identify additional potential legal issues.

DEVELOPMENT OF AVIATION LAW AND REGULATIONS

The predominant theory of airspace rights applied before the advent of aviation derived from the Roman Law maxim *cujus est solum ejus est usque ad coelum*, meaning whoever owns the land possesses all the space above the land extending upwards into the heavens.[7] This maxim was adopted into English common law and eventually made its way into American common law.[8] At the advent of commercial aviation, Congress enacted the Air Commerce Act of 1926[9] and later the 1938 Civil Aeronautics Act.[10] These laws included provisions stating that "to the exclusion of all foreign nations, [the United States has] complete sovereignty of the airspace" over the country.[11] Additionally, Congress declared a "public right of freedom of transit in air commerce through the navigable airspace of the United States."[12] This right to travel in navigable airspace came into conflict with the common law idea that each landowner owned the airspace above the surface in perpetuity. If the common law idea was to be followed faithfully, there could be no right to travel in navigable airspace without constantly trespassing in private property owners' airspace. This conflict was directly addressed by the Supreme Court in *United States v. Causby*, discussed extensively below.

With the passage of the Federal Aviation Act in 1958,[13] the administrator of the FAA was given "full responsibility and authority for the advancement and promulgation of civil aeronautics generally.... "[14] This centralization of responsibility and creation of a uniform set of rules recognized that "aviation is unique among transportation industries in its relation to the federal government—it is the only one whose operations are conducted almost wholly within federal jurisdiction.... "[15] The FAA continues to set uniform rules for the operation of aircraft in the national airspace. In the FAA Modernization and Reform Act of 2012 (FMRA), Congress instructed the FAA to "develop a comprehensive plan to safely accelerate the integration of civil unmanned aircraft systems into the national airspace system."[16] These regulations must provide for this integration "as soon as practicable, but not later than September 30, 2015."[17]

Current FAA Regulations of Navigable Airspace

Fixed-Wing Aircraft

FAA regulations define the minimum safe operating altitudes for different kinds of aircraft. Generally, outside of takeoff and landing, fixed-wing aircraft must be operated at an altitude that allows the aircraft to conduct an emergency landing "without undue hazard to persons or property on the surface."[18] In a congested area, the aircraft must operate at least "1,000 feet above the highest obstacle within a horizontal radius of 2,000 feet of the aircraft."[19] The minimum safe operating altitude over non-congested areas is "500 feet above the surface."[20] Over open water or sparsely populated areas, aircraft "may not be operated closer than 500 feet to any person, vessel, vehicle, or structure."[21] Navigable airspace is defined in statute as the airspace above the minimum safe operating altitudes, including airspace needed for safe takeoff and landing.[22]

Helicopters

While a fixed-wing aircraft is subject to specific minimum safe operating altitudes based on where it is flying, regulation of helicopter minimum altitudes is less rigid. According to FAA regulations, a helicopter may fly below the minimum safe altitudes prescribed for fixed-wing aircraft if it is operated "without hazard to person or property on the surface."[23] Therefore, arguably a helicopter may be lawfully operated outside the zone defined in statute as navigable airspace.[24]

Drones

The FAA does not currently regulate safe minimum operating altitudes for drones as it does for other kinds of aircraft. Defining navigable airspace for drone operation may be one way that the FAA responds to Congress's instruction, in FMRA, to write rules integrating civil drones into the national airspace, which is discussed in more detail below.[25] One possibility is for the FAA to create different classes of drones based on their size and capabilities. Larger drones that physically resemble fixed-wing aircraft could be subject to similar safe minimum operating altitude requirements whereas smaller drones could be regulated similar to helicopters.

Current FAA Regulation of Drones

In 2007, the FAA issued a policy notice stating that "no person may operate a UAS in the National Airspace without specific authority."[26] Therefore, currently all drone operators who do not fall within the recreational use exemption discussed below must apply directly to the FAA for permission to fly.[27]

Public and Civil Operators

Drones operated by federal, state, or local agencies must obtain a certificate of authorization or waiver (COA) from the FAA.[28] After receiving COA applications, the FAA conducts a comprehensive operational and technical review of the drone and can place limits on its operation in order to ensure its safe use in airspace.[29] In response to a directive in FMRA, the FAA recently streamlined the process for obtaining COAs, making it easier to apply on their website.[30] It also employs expedited procedures allowing grants for temporary COAs if needed for time-sensitive missions.[31]

Civil operators, or private commercial operators, must receive a special airworthiness certificate in the experimental category in order to operate.[32] These certificates have been issued on a limited basis for flight tests, demonstrations, and training. Presently, there is no other method of obtaining FAA approval to fly drones for commercial purposes. It appears these restrictions will be loosened in the coming years, since the FAA has been instructed to issue a rulemaking that will lead to the phased-in integration of civilian unmanned aircraft into national airspace.[33]

Recreational Users

The FAA encourages recreational users of model aircraft, which certain types of drones could fall under, to follow a 1981 advisory circular.[34] Under the circular, users are instructed to fly a sufficient distance from populated areas and away from noise-sensitive areas like parks, schools, hospitals, or churches. Additionally, users should not fly in the vicinity of full-scale aircraft or more than 400 feet above the surface. When flying within three miles of an airport, users should notify the air traffic control tower, airport operator, or flight service station. Compliance with these guidelines is voluntary.

Future FAA Regulation of Drones

FMRA instructs the FAA to integrate civil unmanned aircraft systems into the national airspace by the end of FY2015 and implement new standards for public drone operators. This law included provisions describing the comprehensive plan and rulemaking the agency must create to address different aspects of integrating civil drones, restricting the FAA's ability to regulate "model aircraft," and requiring the creation of drone test sites.

Civil Operators

The statute instructs the FAA to create a "comprehensive plan to safely accelerate the integration of civil unmanned aircraft systems into the national airspace" [35] and submit the plan to Congress within one year of enactment.[36] The statute contains a non-exhaustive list of elements that the plan must address, including predictions on how future rulemaking will address the certification process for drones; drone sense and avoid capabilities; and establishing operator or pilot standards, including a licensing and registration system.[37] The plan must also include a timeline for a phased-in approach to integration and ways to ensure the safe operation of civil drones with publicly operated drones in the airspace.[38] The FAA has not yet submitted this comprehensive plan to Congress.

FMRA also directs the FAA to promulgate a series of rules, including rules governing the civil operation of small drones in the national airspace and rules implementing the comprehensive plan described above.[39] Additionally, the FAA must update its 2007 policy statement that established the current scheme of drone authorizations.[40]

Public Operators

As noted above, the FAA has already implemented a streamlined process for public operators to obtain COAs.[41] In addition to this streamlining, FMRA instructs the FAA to "develop and implement operations and certification requirements for the operation of public unmanned aircraft systems in the national airspace."[42] Similar to the provisions governing civil users, these standards must be in place by the end of 2015.

Recreational Users

In FMRA, the FAA was prohibited from promulgating rules regarding certain kinds of model aircraft flown for hobby or recreational use.[43] This prohibition applies if the model aircraft is less than 55 pounds, does not interfere with any manned aircraft, and is flown in accordance with a community-based set of safety guidelines.[44] Additionally, the aircraft must be flown within the line of sight of the operator and be used solely for hobby or recreational purposes.[45] If flown within five miles of an airport, the operator of the model aircraft must notify both the airport operator and air traffic control tower.[46] While the FAA is prohibited from writing rules or regulations governing these aircraft, it is not prohibited from pursuing enforcement actions "against persons operating model aircraft who endanger the safety of the national airspace system."[47]

Test Ranges

As part of its efforts to integrate drones into the national airspace, FMRA also directed the FAA to establish six test ranges that will serve as integration pilot projects.[48] As part of the test range program, the FAA must designate airspace for the operation of both manned and unmanned flights, develop certification and air traffic standards for drones at the test ranges, and coordinate with both NASA and the Department of Defense during development. The test ranges should address both civil and public drone operations.[49]

In February 2013, the FAA published a notice in the Federal Register announcing the process for selection of the sites.[50] In its words, "The overall purpose of this test site program is to develop a body of data and operational experiences to inform integration and the safe operation of these aircraft in the National Airspace System."[51] As directed in the statute, factors for site selection include geographic and climactic diversity and a consideration of the location of the ground infrastructure needed to support the sites.[52] Additionally, in the notice the FAA announced privacy requirements that will be applicable to operations at test sites. These provisions are discussed in more detail below.[53]

The FAA received 50 applications spread across 37 states and is in the process of making its test range site selections.[54]

Airspace and Property Rights

Since the popularization of aviation, courts have had to balance the need for unobstructed air travel and commerce with the rights of private property owners. The foundational case in explaining airspace ownership rights is *United States v. Causby*.[55]

United States v. Causby

In *United States v. Causby*, the Supreme Court directly confronted the question of who owns the airspace above private property.[56] The plaintiffs filed suit against the U. S. government arguing that flights of military planes over their property constituted a violation of the Fifth Amendment Takings Clause, which states that private property shall not "be taken for public use, without just compensation." Generally, takings suits can only be filed against the government when a government actor, as opposed to a private party, causes the alleged harm.[57]

Causby owned a chicken farm outside of Greensboro, North Carolina, that was located near an airport regularly used by the military. The proximity of the airport and the configuration of the farm's structures led the military planes to pass over the property at 83 feet above the surface, which was only 67 feet above the house, 63 feet above the barn, and 18 feet above the tallest tree.[58] While this take-off and landing pattern was conducted according to the Civil Aeronautics Authority guidelines, the planes caused "startling" noises and bright glare at night.

As the Court explained, "as a result of the noise, respondents had to give up their chicken business. As many as six to ten of their chickens were killed in one day by flying into the walls from fright. The total chickens lost in this manner was about 150.... The result was the destruction of the use of the property as a commercial chicken farm."[59] The Court had to determine whether this loss of property constituted a taking without just compensation.

At the outset, the Court directly rejected the common law conception of airspace ownership: "It is ancient doctrine that at common law ownership of the land extended to the periphery of the universe—*Cujus solum ejus est usque ad coelum*. But that doctrine has no place in the modern world."[60] The Court noted that Congress had previously declared a public right of transit in air commerce in navigable airspace and national sovereignty in the airspace.[61] These statutes could not be reconciled with the common law doctrine without subjecting aircraft operators to countless trespass suits. In the Court's words, "common sense revolts at the idea."[62]

Even though it rejected the idea that the *Causbys* held complete ownership of the air up to the heavens, the Court still had to determine if they owned any portion of the space in which the planes flew such that a takings could occur. The government argued that flights within navigable airspace that do not physically invade the surface cannot lead to a taking. It also argued that the landowner does not own any airspace adjacent to the surface "which he has not subjected to possession by the erection of structures or other occupancy."[63]

The Court did not adopt this reasoning, finding instead that "the landowner owns at least as much space above the ground as he can occupy or use in connection with the land. The fact that he does not occupy it in a physical sense—by the erection of building and the like—is not material."[64] Therefore, it found that the landowner owns the airspace in the immediate reaches of the surface necessary to use and enjoy the land and invasions of this space "are in the same category as invasions of the surface."[65] Above these immediate reaches, the airspace is part of the public domain, but the Court declined to draw a clear line. The Court also noted that the government's argument regarding the impossibility of a taking based on flights in navigable airspace was inapplicable in this case because the flights over *Causby*'s land were not within navigable airspace.[66] At the time, federal law defined navigable airspace as the space above the minimum safe flying altitudes for specific areas, but did not include the space needed to take off and land. Even though these flights were not within navigable airspace, the Court seemed to suggest that if they were, the inquiry would not immediately end. Instead, the Court would then have to determine if the regulation itself, defining the navigable airspace, was valid.[67]

Ultimately, in the context of a taking claim, the Court concluded that "flights over private land are not a taking, unless they are so low and so frequent to be a direct and immediate interference with the enjoyment and use of the land."[68] With regard to the *Causbys*' chicken farm, the Court concluded that the military flights had imposed a servitude upon the land, similar to an easement, based on the interference with the use and enjoyment of their property. Although the land did not lose all its economic value, the lower court's findings clearly established the flights led directly to a diminution in the value of the property, since it could no longer be used for its primary purpose as a chicken farm.

Post-Causby Theories of Airspace Ownership

Causby clearly abandoned the ancient idea that private landowners each owned their vertical slice of the airspace above the surface in perpetuity as incompatible with modern life. The case set up three factors to examine in a takings claim that courts still utilize today: (1) whether the planes flew directly over the plaintiff's land; (2) the altitude and frequency of the flights; and (3) whether the flights directly and immediately interfered with the plaintiff's use and enjoyment of the surface land.[69]

However, it left many questions unanswered. Where is the dividing line between the "immediate reaches" of the surface and public domain airspace? Can navigable airspace intersect with the "immediate reaches" belonging to the private property? Can aircraft flying wholly within navigable airspace, as defined by federal law, ever lead to a successful takings claim? How does one assess claims based on lawfully operated aircraft, such as helicopters, flying below navigable airspace?

Subsequent cases have been brought using many different legal claims, including trespass and nuisance, as discussed below, and various ways of describing the resulting injury. Claims could include an "inverse condemnation," another way of describing a taking, or the establishment of an avigation, air, or flying easement. While these legal claims may have different names, it appears that courts use *Causby* as the starting point for analyzing all property-based challenges to intrusions upon airspace. Several different interpretations of *Causby* have emerged in the attempt to articulate an airspace ownership standard, a few of which are described here.

Following *Causby*, several lower courts employed a fixed-height theory and interpreted the decision as creating two distinct categories of airspace. On the one hand, the stratum of airspace that was defined in federal law as "navigable airspace" was always a part of the public domain. Therefore, flights in this navigable airspace could not lead to a successful property-right based action like a takings or trespass claim because the property owner never owned the airspace in the public domain. On the other hand, the airspace below what is defined as navigable airspace could be "owned" by the surface owner and, therefore, intrusions upon it could lead to a successful takings or property tort claim. Since this fixed-height theory of airspace ownership relies heavily on the definition of navigable airspace, the expansion of the federal definition of "navigable airspace" to include the airspace needed to take-off and land[70] greatly impacts what airspace a property owner could claim.

This strict separation between navigable airspace and the airspace a landowner can claim seems to have been disavowed by the Supreme Court. First, in dicta in Braniff Airways v. Nebraska State Board of Equalization & Assessment,[71] a case primarily dealing with the question of federal preemption of state airline regulations, the Court left open the possibility of a taking based on flights occurring in navigable airspace. It summarized *Causby* as holding "that the owner of land might recover for a taking by national use of navigable air space resulting in destruction in whole or in part of the usefulness of the land property."[72] Next, in Griggs v. Allegheny County the Supreme Court found that the low flight of planes over the plaintiff's property, taking off from and landing at a nearby airport's newly constructed runway, constituted a taking that had to be compensated under the Fifth Amendment.[73] The noise and fear of a plane crash caused by the low overhead flights made the property "'undesirable and unbearable'" for residential use, making it impossible for people in the house to converse or sleep.[74] The Court reached this conclusion that a taking occurred based on this injury, despite the fact that the flights were operated properly under federal regulations and never flew outside of navigable airspace.[75] Despite this holding, some lower courts have continued to lend credence to a fixed-height ownership theory as a reasonable interpretation of *Causby*.[76]

Another interpretation of *Causby* essentially creates a presumption of a non-taking when overhead flights occur in navigable airspace. This presumption would recognize the importance of unimpeded travel of air commerce and that Congress placed navigable airspace in the public domain. However, the presumption could be rebutted by evidence that the flights, while in navigable airspace, interfered with the owner's use and enjoyment of the surface enough to justify compensation. As one court reasoned, "as the height of the overflight increases... the Government's interest in maintaining sovereignty becomes weightier while the landowner's interest diminishes, so that the damage showing required increases in a continuum toward showing absolute destruction of all uses of the property."[77]

Finally, some courts have concluded that the altitude of the overhead flight has no determinative impact on whether a taking has occurred. One federal court noted that the government's liability for a taking is not impacted "merely because the flights of Government aircraft are in what Congress has declared to be navigable airspace and subject to its regulation."[78] Under this approach, "although the navigable airspace has been declared to be in the public domain, 'regardless of any congressional limitations, the land owner, as an incident to his ownership, has a claim to the superjacent airspace to the extent that a reasonable use of his land involves such space.'"[79] Under this theory, the court would only need to examine the effect of the overhead flights on the use and enjoyment of the land, and would not need to determine if the flight occurred in navigable airspace.

While the definition of navigable airspace impacts each theory differently, it is clear that under each interpretation a showing of interference with the use and enjoyment of property is required. Cases have clearly established that overhead flights leading to impairment of the owner's livelihood or that cause physical damage qualify as an interference with use and enjoyment of property.[80] Additionally, flights that cause the surface to become impractical for its intended use by the current owner also satisfy the use and enjoyment requirement.[81] For example, in Griggs, the noise, vibration, and fear of damage caused by overhead flights made it impossible for the plaintiffs to converse with others or sleep within their house, leading to their retreat from the property, which had become "undesirable and unbearable for their residential use."[82] Some courts have recognized a reduction in the potential resale value of the property as an interference with its use and enjoyment, even if the property continues to be suitable for the purposes for which it is currently used.[83] One court explained: "Enjoyment of property at common law contemplated the entire bundle of rights and privileges that attached to the ownership of land.... Owners of fee simple estates ... clearly enjoy not only the right to put their land to a particular present use, but also to hold the land for investment and appreciation.... "[84] However, other courts have rejected the idea that restrictions on uses by future inhabitants, without showing loss of property value, are relevant to a determination of the owner's own use and enjoyment of the property.[85]

Trespass and Nuisance Claims against Private Actors

Although *Causby* arose from a Fifth Amendment takings claim, its articulation of airspace ownership standards is also often used in determining state law tort claims such as trespass and nuisance. These state law tort claims could be used to establish liability for overhead flights operated by private actors, where a lack of government involvement precludes a takings claim. Generally, the tort of trespass is any physical intrusion upon property owned by another. However, unlike with surface trespass claims, simply proving that an object or person was physically present in the airspace vertically above the landowner's property is generally not enough to establish a trespass in airspace. Since *Causby* struck down the common law idea of *ad coelum*, landowners generally do not have an absolute possessory right to the airspace above the surface into perpetuity. Instead, airspace trespass claims are often assessed using the same requirements laid out in the *Causby* takings claim. Arguably, these standards are used in property tort claims because there can be no trespass in airspace unless the property owner has some possessory right to the airspace, which was the same question at issue in *Causby*.

To allege an actionable trespass to airspace, the property owner must not only prove that the interference occurred within the immediate reaches of the land, or the airspace that the owner can possess under *Causby*, but also that its presence interferes with the actual use of his land. As one court explained, "a property owner owns only as much air space above his property as he can practicable use. And to constitute an actionable trespass, an intrusion has to be such as to subtract from the owner's use of the property."[86] This standard for airspace trespass was also adopted by the Restatement (Second) of Torts.[87]

Nuisance is a state law tort claim that is not based on possessory rights to property, like trespass, but is rooted in the right to use and enjoy land.[88] Trespass and nuisance claims arising from airspace use are quite similar, since trespass to airspace claims generally require a showing that the object in airspace interfered with use and enjoyment of land. However, unlike trespass, nuisance claims do not require a showing that the interference actually occupied the owner's airspace. Instead, a nuisance claim can succeed even if the interference flew over adjoining lands and never directly over the plaintiff's land, as long as the flight constitutes a substantial and unreasonable interference with the use and enjoyment of the land.

Potential Liability Arising from Civilian Drone Use

The integration of drones into domestic airspace will raise novel questions of how to apply existing airspace ownership law to this new technology. How courts may apply the various interpretations of *Causby*, discussed above, to drones will likely be greatly impacted by the FAA's definition of navigable airspace for drones.

The potential for successful takings, trespass, or nuisance claims from drone use will also be impacted by the physical characteristics of the drone, especially given that current case law heavily emphasizes the impact of the flight on use and enjoyment of the surface property. Several characteristics of drones may make their operation in airspace less likely to lead to liability for drone operators than for aircraft operators. First, the noise attributed to drone use may be significantly less than noise created by helicopters or planes powered by jet engines. Second, drones commonly used for civilian purposes could be much smaller than common aircraft used today. This decreased size is likely to lead to fewer physical impacts upon surface land such as vibration and dust, which are common complaints arising from overhead aircraft and helicopter flights. Finally, it is unknown at this time how most drones will be deployed into flight. Will drone "airports" be used to launch the aircraft or will they take off and land primarily from individual property? If drone use remains decentralized and is not organized around an "airport," then drones are less likely to fly repeatedly over the same piece of property, creating fewer potential takings, trespass, or nuisance claims. Additionally, the majority of drones are more likely to operate like helicopters, taking off and landing vertically, than like traditional fixed-wing aircraft. This method of takeoff reduces the amount of surface the aircraft would have to fly over before reaching its desired flying altitude, minimizing the potential number of property owners alleging physical invasion of the immediate reaches of their surface property.

Alternatively, the potential ability for drones to fly safely at much lower altitudes than fixed-wing aircraft or helicopters could lead to a larger number of property-based claims. Low-flying drones are more likely to invade the immediate reaches of the surface property, thus satisfying part of the requirement for a takings or trespass claim.

PRIVACY

Perhaps the most contentious issue concerning the introduction of drones into U. S. airspace is the threat that this technology will be used to spy on American citizens. With the ability to house high-powered cameras, infrared sensors, facial recognition technology, and license plate readers, some argue that drones present a substantial privacy risk.[89] Undoubtedly, the government's use of drones for domestic surveillance operations implicates the Fourth Amendment and other applicable laws.[90] In like manner, privacy advocates have warned that private actors might use drones in a way that could infringe upon fundamental privacy rights.[91] This section will focus on the privacy issues associated with the use of drones by private, non-governmental actors. It will provide a general history of privacy law in the United States and survey the various privacy torts, including intrusion upon seclusion, the privacy tort most applicable to drone surveillance. It will then explore the First Amendment right to gather news. Application of these theories to drone surveillance will be discussed in the section titled "Congressional Response."

Early Privacy Jurisprudence

Although early Anglo-Saxon law lacked express privacy protections, property law and trespass theories served as proxy for the protection of individual privacy. Lord Coke pronounced in 1605 that "the house of everyone is to him as his castle and fortress, as well for his defence against injury and violence, as for his respose[.]"[92] This proposition that individuals are entitled to privacy while in their homes crossed the Atlantic with the colonists and appeared prominently in early revolutionary thinking.[93] In one early American common law decision, the court noted that "[t]he law is clearly settled, that an officer cannot justify the breaking open an outward door or window, in order to execute process in a civil suit; if he doth, he is a trespasser."[94] In cases lacking physical trespass, prosecutors relied on an eavesdropping theory, which protected the privacy of individuals' conversations while in their home.[95]

These century-old theories of trespass and eavesdropping, however, failed to keep up with a rapidly changing society fueled by advancing technologies. As with today's celebrity-obsessed society, late-19th century society experienced the birth and spread of "yellow journalism," a new media aimed at emphasizing the "curious, dramatic, and unusual, providing readers a 'palliative of sin, sex, and violence.'"[96] Faster presses and instantaneous photography enabled journalists to exploit and spread gossip.[97] Louis D. Brandeis (then a private attorney) and Samuel Warren were bothered with the press's constant intrusions into the private affairs of prominent Bostonians.[98] In 1890, they published a seminal law review article formulating a new legal theory–the right to be let alone.[99] Brandeis and Warren understood that existing tort doctrines such as trespass and libel were insufficient to protect privacy rights, as "only a part of the pain, pleasure, and profit of life lay in physical things."[100] They noted that this new right to privacy derived not from "the principle of private property, but that of an inviolate personality."[101] The authors observed that "instantaneous photographs and newspaper enterprise have invaded the sacred precincts of private and domestic life; and numerous mechanical devices threaten to make good the prediction that 'what is whispered in the closet shall be proclaimed from the house-tops.'"[102] Although this new theory had its detractors,[103] it found its way into the common law of several states.[104]

Privacy Torts

In 1939, the First Restatement of Torts (a set of model rules intended for adoption by the states) created a general tort for invasion of privacy.[105] By 1940, a minority of states had adopted some right of privacy either by statute or judicial decision, and six states had expressly refused to adopt such a right.[106] Twenty years later, Dean William Prosser surveyed the case law surrounding this right and concluded that the right to privacy entailed four distinct (yet, sometimes overlapping) rights: (1) intrusion upon seclusion; (2) public disclosure of private facts; (3) publicity which puts the target in a false light; and (4) appropriation of one's likeness.[107] These four categories were incorporated into the Restatement (Second) of Torts.[108]

Section 652B of the Restatement (Second) of Torts creates a cause of action for intrusion upon seclusion,[109] the privacy tort most likely to apply to drone surveillance.[110] It has been adopted either by common law or statute in an overwhelming majority of the states.[111] Section 652B provides: "One who intentionally intrudes, physically or otherwise, upon the solitude or seclusion of another or his private affairs or concerns, is subject to liability to the other for invasion of his privacy, if the intrusion would be highly offensive to a reasonable person."[112] Courts have developed a set of rules for applying Section 652B. First, it requires an objective person standard, testing whether a person of "ordinary sensibilities" would be offended by the alleged invasion.[113] Thus, someone with an idiosyncratic sensitivity—say, an aversion to cameras—could not satisfy this standard by simply having his photograph taken. Likewise, the intrusion must not only be offensive, but "*highly* offensive,"[114] or as one court put it, "outrageously unreasonable conduct."[115] Generally, a single incident will not suffice; instead, the intrusion must be "repeated with such persistence and frequency as to amount to a course of hounding" and "becomes a burden to his existence.... "[116] However, in a few cases a single intrusion was adequate.[117] The invasion of privacy must been intentional, meaning the defendant must desire that the intrusion would occur, or as with other torts,[118] knew with a substantial certainty that such an invasion would result from his actions.[119] An accidental intrusion is not actionable. Finally, in some states, the intrusion must cause mental suffering, shame, or humiliation to permit recovery.[120]

A review of the case law demonstrates that the location of the target of the surveillance is, in many cases, determinative of whether someone has a viable claim for intrusion upon seclusion. For the most part, conducting surveillance of a person while within the confines of his home will constitute an intrusion upon seclusion.[121] The illustrations to Section 652B offer an example of a private detective who photographs an individual while in his home with a telescopic camera as a viable claim.[122] Likewise, as one court observed, "when a picture is taken of a plaintiff while he is in the privacy of his home, ... the taking of the picture may be considered an intrusion into the plaintiff's privacy just as eavesdropping or looking into his upstairs windows with binoculars are considered an invasion of his privacy."[123]

The likelihood of a successful claim is diminished if the surveillance is conducted in a public place. The comments to Section 652B explain that there is generally no liability for photographing or observing a person while in public "since he is not then in seclusion, and his appearance is public and open to the public eye."[124] Likewise, Prosser observed:

> On the public street, or in any other public place, the plaintiff has no right to be alone, and it is no invasion of his privacy to do no more than follow him about. Neither is it such an invasion to take a photograph in such a place, since this amounts to nothing more than making a record, not differing essentially from a full written description, of a public sight which anyone present would be free to see.[125]

The case law also supports this proposition. The Alabama Supreme Court dismissed a claim of wrongful intrusion against operators of a race track who photographed the plaintiffs while they were in the "winner's circle" at the track.[126] Similarly, a federal district court dismissed a claim by a husband and wife who had been photographed by Forbes Magazine while waiting in line at the Miami International Airport as it was taken in "a place open to the general public."[127] Likewise, a Vietnam veteran lost a claim for invasion of privacy based on photographs that depicted him and other soldiers during a combat mission in Vietnam–again, a public setting.[128] Other examples include the recording of license plate numbers of cars parked in a public parking lot[129] and photographing a person while walking on a public sidewalk.[130]

Indeed, even plaintiffs who were videotaped or photographed while on their own property have generally been unsuccessful in their privacy claims so long as they could be viewed from a public vantage point. Rejecting one plaintiff's claim for intrusion upon seclusion, the Supreme Court of Oregon held that even though the investigators trespassed on the plaintiff's property to film him, the investigation did not "constitute an unreasonable surveillance 'highly offensive to a reasonable man[,]'"[131] as the plaintiff could have been viewed from the road by his neighbors or passersby.[132] In another case, the wife of a prominent Puerto Rican politician sought damages from a newspaper for invasion of privacy allegedly committed when an agent of the newspaper photographed her house as part of a news story about her husband.[133] The court dismissed her claim as the photographers were not "unreasonably intrusive," and the photographs depicted only the outside of the home and no persons were photographed.[134] Similarly, in one case a couple sued a cell phone company for intrusion upon seclusion when the company's workers looked onto their property each time they serviced a nearby cell tower.[135] The court rejected their claim, holding that "[t]he mere fact that maintenance workers come to an adjoining property as part of their work and look over into the adjoining yard is legally insufficient evidence of highly offensive conduct."[136] There are many other examples.[137]

However, there have been some successful claims for intrusion upon seclusion involving surveillance conducted in public.[138] The comments to Section 652B explain: "Even in a public place, however, there may be some matters about the plaintiff, such as his underwear or lack of it, that are not exhibited to the public gaze, and there may still be invasion of privacy when there is intrusion upon these matters."[139] One of the most famous cases concerning this "public gaze" theory involved a suit for invasion of privacy against a newspaper when it published a picture of the plaintiff with her dress blown up as she was leaving a fun house at a county fair.[140] In upholding the plaintiff's claim, the court observed: "To hold that one who is involuntarily and instantaneously enmeshed in an embarrassing pose forfeits her right of privacy merely because she happened at the moment to be part of a public scene would be illogical, wrong, and unjust."[141] In *Huskey v. National Broadcasting Co. Inc.*, a prisoner sued NBC, a television broadcasting company, alleging that by filming him without consent while he was working out in the exercise yard at the prison, NBC invaded his privacy.[142] NBC countered that depictions of persons in a "publicly visible area" could not support the claim for invasion of seclusion.[143] Ultimately, the court permitted the prisoner's claim to go forward, observing that "[o]f course [the prisoner] could be seen by guards, prison personnel and inmates, and obviously he was in fact seen by NBC's camera operator. But the mere fact a person can be seen by others does not mean that person cannot legally be 'secluded.'"[144] Although relief is available for certain cases of public surveillance, recovery seems to be the exception rather than the norm.[145]

First Amendment and Newsgathering Activities

Based on the foregoing discussion, safeguarding privacy from intrusive drone surveillance is clearly an important societal interest. However, this interest must be weighed against the public's countervailing concern in securing the free flow of information that inevitably feeds the "free trade of ideas."[146] Unmanned aircraft can improve the press and the public's ability to gather news: they can operate in dangerous areas without putting a human operator at risk of danger; can carry sophisticated surveillance technology; can fly in areas not currently accessible by traditional aircraft; and can stay in flight for long durations. However, challenges arise in attempting to find an appropriate balance between this interest in newsgathering and the competing privacy interests at stake.

The First Amendment to the United States Constitution provides that "Congress shall make no law ... abridging the freedom of speech, or of the press.... "[147] The Court has construed this phrase to cover not only traditional forms of speech, such as political speeches or polemical articles, but also conduct that is "necessary for, or integrally tied to, acts of expression,"[148] such as distribution of political literature[149] or door-to-door solicitation.[150] Additionally, the Court has pulled within the First Amendment's protection other conduct that is not expressive in itself, but is "necessary to accord full meaning and substance to those guarantees."[151] For example, the Court has said that the public is entitled to a "right to receive news" as a correlative of the right to free expression.[152]

Like this right to receive news, the Court has intimated in a series of cases beginning in the 1960s that the public and the press may be entitled to *a right to gather news* under the First Amendment. Initially, in *Zemel v. Rusk*, the Court observed that the right "to speak and publish does not carry with it the unrestrained right to gather information."[153] The Court's reluctance to extend this right may have signaled its concern that an unconditional newsgathering right could subsume almost any government regulation that places a slight restriction on the ability to gather news.[154] However, several years later the Court indicated in *Branzburg v. Hayes* that although laws of general applicability apply equally to the press as to the general public, that "[n]ews gathering is not without its First Amendment protections,"[155] and that "without some protection for seeking out the news, freedom of the press could be eviscerated."[156] The Court, however, failed to clearly delineate the parameters of such a protection. In the Court's most recent case, *Cohen v. Cowles Media Co.*, the Court adhered to the "well-established line of decisions holding that generally applicable laws do not offend the First Amendment simply because their enforcement against the press has incidental effects on its ability to gather and report the news."[157] The Court noted that it is "beyond dispute 'that the publisher of a newspaper has no special immunity from the application of general laws. He has no special privilege to invade the rights of others.'"[158]

The lower federal courts have explored this right to gather news in the context of photographing or video recording. In *Dietemann v. Time, Inc.* the Ninth Circuit Court of Appeals explored the extent to which reporters could use surreptitious means to carry out their newsgathering.[159] There, defendants Time Life sent undercover reporters to a man's house where he claimed to use minerals and other materials to heal the sick. The reporters used a hidden camera to take pictures of the man, and a hidden microphone to transmit the conversation to other operatives. The defendants claimed that the First Amendment's right to freedom of the press shielded its newsgathering activities. In rejecting this claim, the court observed that although an individual accepts the risk when inviting a person into his home that the visitor may repeat the conversation to a third party, "he does not and should not be required to take the risk that what is heard and seen will be transmitted by photograph or recording, or in our modern world, in full living color and hi-fi to the public at large or to any segment of it that the visitor may select."[160] The court held that "hidden mechanical contrivances" are not indispensable tools of investigative reporting, and that the "First Amendment has never been construed to accord newsman immunity from torts or crimes committed during the course of newsgathering."[161] In *Galella v. Onassis*, Galella, a self-proclaimed "paparazzo," constantly followed around, harassed, and photographed Jacqueline Kennedy Onassis and her children.[162] As part of an ongoing lawsuit, Onassis sued Galella for, *inter alia*, invasion of her and her family's privacy. Galella argued that he was entitled to the absolute "wall of immunity" that protects newsmen under the First Amendment. The Second Circuit Court of Appeals quickly rejected this absolutist position: "There is no such scope to the First Amendment right. Crimes and torts committed in news gathering are not protected. There is no threat to a free press in requiring its agents to act within the law."[163] By contrast, the Seventh Circuit in *Desnick v. American Broadcast Companies, Inc.* held that surreptitious recording was not a privacy invasion because the target of the surveillance was a party to the conversation, thereby vitiating any claim to privacy in those conversations.[164]

Congressional Response

If Congress chooses to act, it could create privacy protections to protect individuals from intrusive drone surveillance conducted by private actors.[165] Such proposals would be considered in the context of the First Amendment rights to gather and receive news. Several bills have been introduced in the 113th Congress that would regulate the private use of drones. Additionally, there are other measures Congress could adopt.

Drone Aircraft Privacy and Transparency Act of 2013 (H.R. 1262)

In the 113th Congress, Representative Ed Markey introduced the Drone Aircraft Privacy and Transparency Act of 2013 (H.R. 1262).[166] This bill would amend FMRA to create a comprehensive scheme to regulate the private use of drones, including data collection requirements and enforcement mechanisms. First, this bill would require the Secretary of Transportation, with input from the Secretary of Commerce, the Chairman of the Federal Trade Commission, and the Chief Privacy Officer of the Department of Homeland Security, to study any potential threats to privacy protections posed by the introduction of drones in the national airspace. Next, the bill would prohibit the FAA from issuing a license to operate a drone unless the application for such use included a "data collection statement." This statement would require the following items: a list of individuals who would have the authority to operate the drone; the location in which the drone will be used; the maximum period it will be used; and whether the drone would be collecting information about individuals. If the drone will be used to collect personal information, the statement must include the circumstances in which such information will be used; the kinds of information collected and the conclusions drawn from it; the type of data minimization procedures to be employed; whether the information will be sold, and if so, under what circumstances; how long the information would be stored; and procedures for destroying irrelevant data. The statement must also include information about the possible impact on privacy protections posed by the operation under that license and steps to be taken to mitigate this impact. Additionally, the statement must include the contact information of the drone operator; a process for determining what information has been collected about an individual; and a process for challenging the accuracy of such data.

Finally, the FAA would be required to post the data collection statement on the Internet.

H.R. 1262 includes several enforcement mechanisms. First, the FAA may revoke any license of a user that does not comply with these requirements. The Federal Trade Commission would have the primary authority to enforce the data collection requirements just stated. Additionally, the Attorney General of each state, or an official or agency of a state, is empowered to file a civil suit if there is reason to believe that the privacy interests of residents of that state have been threatened or adversely affected. H.R. 1262 would also create a private right of action for a person injured by a violation of this legislation.

Preserving American Privacy Act of 2013 (H.R. 637)

Representative Poe introduced the Preserving American Act of 2013 (H.R. 637) which would prohibit the use of drones to capture images in a manner highly offensive to a reasonable person where the person is engaging in a personal or familial activity under circumstances in which the individual has a reasonable expectation of privacy, regardless of whether there is a physical trespass.[167]

Other Proposals

Additionally, Congress could create a cause of action for surveillance conducted by drones similar to the intrusion upon seclusion tort provided under Restatement Section 652B.[168] How would a court assess whether drone surveillance violated this type of tort? First, generally speaking, the location of the search would be determinative of whether a person is entitled to an expectation of privacy. Although courts have posited that the common law, like the Fourth Amendment, is intended to "protect people, not places[,]"[169] the *location* of an alleged intrusion factors heavily in a privacy analysis. The greatest chance for liability occurs when a person photographs or videotapes another while in the seclusion of his home. While technology has increasingly shrunk other spheres of privacy in the digital age, the home is still accorded significant legal protection. Using a drone to peer inside the home of another—whether looking through a window or utilizing extra-sensory technology such as thermal imaging—would likely constitute an intrusion upon seclusion. Moving from the home to a public space, or even a space on private property where one can be seen from a public vantage point, significantly reduces the chance of tort liability. However, certain instances of highly offensive surveillance in public may be actionable.

This leads to the second factor that will inform a reviewing court's analysis: the degree of offensiveness of the surveillance. The Ninth Circuit Court of Appeals, applying California law, observed that, in determining offensiveness, "common law courts consider, among other things: 'the degree of intrusion, the context, conduct and circumstances surrounding the intrusion as well as the intruder's motives and objectives, the setting into which he intrudes, and the expectations of those whose privacy is invaded.'" Several of these factors—especially, the context of the intrusion and the motive of the intruder—are fact intensive and require application in a particular case to fully understand. However, some generalizations can be made. The cases discussed above that did find an intrusion upon seclusion in a public place required highly offensive activity, such as closely following another person for an extended period or photographing another in a highly embarrassing shot. Likewise, a court might recognize liability if one were to use a drone to follow another for an extended period of time, particularly at a close distance. It is not clear, however, whether knowledge of being surveilled makes the monitoring more or less offensive. For example, one court seemed to rely on the fact that the defendant was unaware that her house was being photographed to hold that she did not have a viable privacy claim.[170] A drone flying at several thousand feet may not significantly disturb the target of the surveillance and could fall within this rationale. Nevertheless, filming someone in a compromising or embarrassing situation without his knowledge can be equally offensive. Here, the facts of the particular case would determine liability.

Congress could also create a privacy statute tailored to drone use similar to the anti-voyeurism statutes, or "Peeping Tom" laws, enacted in many states.[171] These laws prohibit persons from surreptitiously filming others in various circumstances and places.[172] Some states prohibit surreptitious surveillance of a person while on private property, usually a private residence.[173] Nevada employs this model, prohibiting a person from entering the property of another with the intent to peep through a window of the building.[174] Likewise, New Jersey prohibits a person from peering into the window of the dwelling of another "under circumstances in which a reasonable person in the dwelling would not expect to be observed."[175] Other states require a prurient intent when conducting the surveillance. Under Washington State's statute, a person commits the crime of voyeurism if, for the purpose of arousing or gratifying his sexual desire, he films or photographs (1) a person in a place where he or she would expect privacy; or (2) the intimate areas of another person, whether he or she is in a public or private place.[176]

Similarly, Congress could adopt an "anti-paparazzi" statute, like that enacted in California, to prevent intrusive drone surveillance.[177] In fact, Congress considered a similar measure in the 105th Congress. The Privacy Protection Act of 1998 and the Personal Intrusion Act of 1998 would have made it unlawful to persistently follow or chase another person for the purpose of obtaining a visual image of that person if the plaintiff met the following elements: (1) the image was transferred in interstate commerce or the person taking the photograph traveled in interstate commerce; (2) the person had a reasonable expectation of privacy from such intrusion; (3) the person feared death or bodily injury from being chased; and (4) the taking of the image was for commercial purposes.[178] Also, these bills would have created a civil remedy for an individual whose privacy was intruded upon. Congress could use this model to make it unlawful to persistently monitor another person using drone surveillance.

FAA Regulation of Privacy

Some observers have questioned whether the FAA has the legal authority to create privacy protections as it begins to integrate drones in the national airspace.[179] This section will explore the FAA's legal authority to establish privacy protections when it engages in rulemaking and establishes the six drone test ranges as required under FMRA.

It is well settled that agencies do not wield inherent powers, and that any authority they do have must be delegated by Congress.[180] Thus, when engaging in rulemaking or any other administrative action, the agency must be able to identify a specific statutory source of authority. In *Chevron v. Natural Resources Defense Council*, the Supreme Court established a two-part test (now known as the *Chevron* two step) that assesses whether a federal agency should be accorded deference in interpreting and implementing its authorizing statute or a statute it administers.[181] First, this test asks "whether Congress has directly spoken to the precise question at issue."[182] If so, the analysis ends there and the court and the agency "must give effect to the unambiguously expressed intent of Congress."[183] If, however, "the statute is silent or ambiguous" the court must then determine if the agency's interpretation is a "permissible construction of the statute."[184] This type of *Chevron* analysis may be applied by a reviewing court if the FAA promulgated rules governing privacy, as part of the FMRA directed rulemaking and those rules were challenged.

Rulemaking Required by FMRA

FMRA directs the FAA to engage in two sets of rulemaking. The first rulemaking requires that by August 14, 2014, the FAA issue a final rule on integrating "small unmanned aircraft systems" into the national airspace system.[185] The second rulemaking requires the FAA to develop a "comprehensive plan to safely accelerate the integration of civil unmanned aircraft systems into the national airspace system[;]"[186] provide notice on proposed rulemaking to implement this comprehensive plan not later than August 14, 2014; and publish a final rule by December 14, 2015.[187]

As to the small drone rulemaking, FMRA provides little guidance on what factors should inform the agency's rulemaking. The act merely requires that the FAA issue a final rule "on small unmanned aircraft systems that will allow for civil operation of such systems in the national airspace system.... "[188] Applying the first step of *Chevron*, it is clear that Congress did not address the precise issue at question–that is, Congress did not expressly provide FAA authority to regulate privacy. Accordingly, a reviewing court would then have to assess under step two whether addressing privacy as part of rulemaking would be a reasonable interpretation of FMRA. There are plausible arguments on both sides of this question.

The Court asserted in *Chevron* that "the power of an administrative agency to administer a congressionally created ... program necessarily requires the formulation of policy and the making of rules to fill any gap left, implicitly or explicitly, by Congress."[189] Given that privacy has been a paramount concern of both the public and various Members of Congress, it would seem odd if Congress delegated the FAA rulemaking authority to integrate drones into the national airspace, but withheld the authority to regulate one of the most prominent and controversial issues surrounding this integration. Rather, one could argue that Congress would have inferred that the FAA would fill in these gaps with reasonable regulations, including those regulating privacy and data collection issues. Conversely, those who contend that the FAA has neither the legal authority nor the expertise to regulate privacy issues concerning drone use may point to the FAA's organic statute. This law provides the FAA authority to ensure the safety and efficiency of air travel,[190] but appears to contain no *express* authority to regulate privacy. They may further argue that the FAA has not historically regulated privacy as it pertains to persons or things on the ground in relation to traditional air flight, and currently does not have the technical expertise to undertake such regulations. These arguments could support the theory that Congress intentionally omitted privacy regulation from the FAA's purview when conducting this required rulemaking.

Next, as to the comprehensive plan rulemaking, Congress has provided some guidance as to the factors the FAA should take into consideration, but none of the factors discuss privacy concerns.[191] Thus, like the rulemaking for small drones, under the *Chevron* first step, Congress has not spoken directly to the issue in question. Moving to the second step, would it be reasonable for the FAA to include privacy regulations in its rulemaking implementing this comprehensive plan? First, the use of the term "at a minimum" as a preface to the list of factors to be considered in this comprehensive plan and rulemaking make it illustrative, not exhaustive. This phrasing arguably suggests that Congress understood that the FAA might address other factors, perhaps including privacy, beyond those enumerated in section 332. Second, section 332 provides that the FAA must "define the acceptable standards for operation and certification of civil unmanned aircraft systems."[192] Viewing this language in light of *Chevron* deference, a court could find that regulating requirements that protect privacy fall within the "acceptable standards for operation" of drones in the national airspace.

In sum, it appears that the open-ended nature of Congress's instructions to the FAA, coupled with the prominence of privacy concerns, would likely persuade a court that the FAA's potential regulation of privacy as part of formal rulemaking is a reasonable interpretation of FMRA that should be accorded deference under a *Chevron* analysis.

Test Ranges and Privacy

In addition to the rulemaking described above, section 332(c) of FMRA requires the FAA Administrator to "establish a program to integrate unmanned aircraft systems into the national airspace system at 6 test ranges."[193] On February 22, 2013, the FAA issued a request for comment on the privacy rules that will apply to test range operators.[194] In its request for comment, the FAA proposed several requirements that might apply to the operation of these test ranges. [195] Once the FAA selects the site operators, each must enter into an Other Transaction Agreement (OTA) with the FAA–a legally binding agreement setting out the terms and conditions under which the site will be operated. This request for comment is intended to provide the public the ability to comment on "potential privacy considerations, associated reporting requirements, and how the FAA can help ensure privacy considerations are addressed through mechanisms put in place as a result of the OTA."[196]

This FAA announcement raises another legal question: does the FAA have the authority to regulate privacy via OTA agreements entered into with the test range operators? As a threshold issue, it is not clear what level of deference a court would apply to this administrative action. In certain instances, agency actions that do not amount to formal rulemaking have not been accorded *Chevron* deference. In *Christensen v. Harris County*, the Supreme Court held that a Department of Labor opinion letter interpreting the Family Medical Leave Act was not entitled to deference under *Chevron*.[197] The Court observed that "[i]nterpretations such as those in opinion letters—like interpretations contained in policy statements, agency manuals, and enforcement guidelines, all of which lack the force of law—do not warrant *Chevron*-style deference."[198] Instead, interpretations contained in administrative pronouncements such as opinion letters are entitled to some deference under the rule pronounced in *Skidmore v. Swift & Co.*,[199] "but only to the extent that those interpretations have the 'power to persuade.'"[200] In *United States v. Mead*, the Court again ruled that *Skidmore*, not *Chevron*, deference applied to a United States Custom Service opinion letter setting tariff levels on certain imports.[201]

A reviewing court could apply the *Christensen-Mead* line of cases to hold that the lower level deference accorded under *Skidmore* should apply to the FAA's use of the OTAs in establishing the test ranges. As in those cases, the OTAs would not have the force of law and would not be the product of formal agency adjudication or rulemaking. These factors weigh against applying *Chevron*'s deferential approach.

However, *Mead* suggested that *Chevron* deference may be due when the agency conducts notice and comment procedures as part of its interpretive process, which were not utilized in either *Christensen* or *Mead*.[202] Here, the FAA has issued a notice for comment on the proposed privacy regulations that will be included in the OTAs. This fact might persuade a court into applying the more deferential *Chevron* test.

Under either level of scrutiny, it is not at all clear whether the FAA would have the authority to regulate privacy as part of the OTAs. Congress did not speak to this issue in FMRA.[203] Thus, a reviewing court would have to determine if the agency's regulation of privacy is either a reasonable interpretation of the statute under *Chevron* or has the "power to persuade" under *Skidmore*. Some of the same factors that arguably support the inclusion of privacy in the formal rulemaking could apply equally to the test ranges. The idea that Congress left it to the FAA to fill in the gaps in establishing the test ranges, and that privacy is one of the primary concerns surrounding the integration of drones into U. S. airspace, could be offered as an argument to uphold the FAA's regulation of privacy. On the other side of the ledger, the act's enumerated list of factors to be addressed at these test ranges is primarily focused on safety issues and does not expressly permit the FAA to regulate privacy. One could argue that this formulation evinces Congress's intent for the FAA to focus on safety, the FAA's stock and trade, rather than privacy, an area in which the FAA appears to have little experience.

RELATED LEGAL ISSUES

In addition to the legal issues described above, there are a host of other issues that may arise when introducing drones into the U. S. national airspace system.

Preemption of State and Local Regulations. The increased presence of drones in domestic airspace raises the question of potential federal preemption of state or local efforts to regulate different aspects of drone use. The doctrine of preemption derives from the Supremacy Clause of the Constitution, which states that federal law, treaties, and the Constitution are the "supreme Law of the Land."[204] A federal law may preempt state or local action in one of three ways: if the statute expressly states its intent to preempt state or local action (express preemption); if a court concludes that Congress intended to occupy the regulatory field, implicitly preventing state or local action in that area (field preemption); or if the state or local action directly conflicts with or frustrates the purpose of the federal provisions (conflict preemption).[205]

With regard to traditional aviation laws, generally, state regulations of aviation safety, airspace management, and aviation noise are preempted by federal laws and regulations.[206] Congress vested sole responsibility for the aviation industry and domestic airspace with the federal government in the Federal Aviation Act of 1958.[207] According to the legislative history, the FAA was to have "full responsibility and authority for the advancement and promulgation of civil aeronautics generally, including promulgation and enforcement of safety regulations."[208] In *City of Burbank v. Lockheed Air Terminal, Inc.*, the Supreme Court struck down a local city ordinance that prohibited planes from taking off during certain hours of the day as preempted by the federal regulatory scheme.[209] Expressing its fear regarding local control of airspace, the Court stated, "If we were to uphold the Burbank ordinance and a significant number of municipalities followed suit, it is obvious that fractionalized control of the timing of takeoffs and landings would severely limit the flexibility of the FAA in controlling air traffic flow."[210] The Supreme Court has, however, upheld state regulations imposing taxes on aircraft equipment located within the state.[211]

State proposals seeking to regulate the use of drones are currently pending in many state legislatures throughout the country.[212] The Virginia General Assembly has passed a two-year moratorium on the use of drones by state and local law enforcement.[213] The bill prohibits the use of drones by agencies with jurisdiction over criminal law enforcement or regulatory violations, but includes exceptions for emergency situations. Following passage of the bill, the Governor neither signed nor vetoed the bill, but rather sent it back to the General Assembly with amendments, where it now awaits further action. Several other states have introduced bills similarly targeting the use of drones for surveillance.[214] Other states, like Texas, have introduced bills attempting to address privacy concerns related to widespread drone use. The Texas proposal would create a new state misdemeanor when a person uses a drone to capture an image without the consent of the landowner who owns the property captured in the image.[215]

If these proposals were implemented, questions about federal preemption may be raised. It appears that field preemption or conflict preemption would be the most likely grounds for finding preemption of such state regulations based on current federal law, if at all, since FMRA does not contain an express preemption clause. The extent to which the state can regulate drone use without being preempted by federal law may depend on the scope of the forthcoming federal regulations, the nature of the state regulations, and a reviewing court's analysis of whether Congress intended to "occupy the field" of regulation on that issue. The Court has determined that field preemption can be inferred when "the pervasiveness of the federal regulation precludes supplementation by the States, where the federal interest in the field is sufficiently dominant, or where the object sought to be obtained by the federal law and the character of obligations imposed by it reveal the same purpose."[216]

Right to Protect Property from Trespassing Drones. There may be instances where a landowner is entitled to protect his property from intrusion by a drone. Under Restatement (Second) of Torts Section 260, "one is privileged to commit an act which would otherwise be a trespass to a chattel or a conversion if the act is, or is reasonably believed to be, necessary to protect the actor's land or chattels or his possession of them, and the harm inflicted is not unreasonable as compared with the harm threatened."[217] What this means is, in certain instances, a landowner would not be liable to the owner of a drone for damage necessarily or accidentally resulting from removing it from his property. However, there appear to be no cases where a landowner was permitted to use force to prevent or remove an aircraft from his property.

Additionally, as discussed above, determining whether a drone in flight is trespassing upon one's property may be unusually challenging.

Stalking, Harassment, and Other Crimes. Traditional crimes such as stalking, harassment, voyeurism, and wiretapping may all be committed through the operation of a drone. As drones are further introduced into the national airspace, courts will have to work this new form of technology into their jurisprudence, and legislatures might amend these various statutes to expressly include crimes committed with a drone.

Wiretap Laws. Under the federal wiretap statute, it is unlawful to intentionally intercept an "oral communication"[218] by a person "exhibiting an expectation that such communication is not subject to interception under circumstances justifying such expectation.... "[219] Currently, commercial microphones can record sounds upwards of 300 feet.[220] Use of such a microphone on a drone to record private conversations could implicate the federal wiretap statute.

CONCLUSION

The legal issues discussed in this report will likely remain unresolved until the civilian use of drones becomes more widespread. To that end, the FAA has been tasked with developing "a comprehensive plan to safely accelerate the integration" of drones into the national airspace, which focuses on the safety of the drone technology and operator certification. While the deadline for development of the plan has already elapsed, the FAA has until the end of FY2015 to implement such a plan.[221] Additionally, the FAA must identify six test ranges where it will integrate drones into the national airspace. This deadline, 180 days after enactment of the act, has also elapsed without FAA compliance. Once these regulations are tested and promulgated, the unique legal challenges that could arise based on the operational differences between drones and already ubiquitous fixed-wing aircraft and helicopters may come into sharper focus.

End Notes

[1] A "drone" is simply an aircraft that can fly without a human operator. They are sometimes referred to as unmanned aerial vehicles (UAV), and the whole system—including the aircraft, the operator on the ground, and the digital network required to fly the aircraft—is

referred to as an unmanned aircraft system (UAS). *See generally* CRS Report R42718, *Pilotless Drones: Background and Considerations for Congress Regarding Unmanned Aircraft Operations in the National Airspace System*, by Bart Elias.

[2] *See* GOV'T ACCOUNTABILITY OFFICE, UNMANNED AIRCRAFT SYSTEMS: MEASURING PROGRESS AND ADDRESSING POTENTIAL PRIVACY CONCERNS WOULD FACILITATE INTEGRATION INTO THE NATIONAL AIRSPACE SYSTEM (2012).

[3] It is reported that News Corp. has used a small drone to monitor storm damage in Alabama and flooding in North Dakota. Kashmir Hill, *FAA Looks Into News Corp's Daily Drone, Raising Questions About Who Gets to Fly Drones in the U. S.*, FORBES, (August 2, 2011 3:52 P.M.), http://www.forbes.com/sites/kashmirhill/2011/08/02/faa-looks-intonews-corps-daily-drone-raising-questions-about-who-gets-to-fly-drones-in-the-u-s/.

[4] Nick Wingfield & Somini Sengupta, *Drones Set Sights on U. S. Skies*, N.Y. TIMES (February 17, 2012), *available at* http://www.nytimes.com/2012/02/18/technology all&_r=0.

[5] FAA Modernization and Reform Act of 2012, P.L. 112-95, 126 Stat. 11.

[6] Groups such as the Association for Unmanned Vehicle Systems International, which boasts 7,200 members, including defense contractors, educational institutions, and government agencies, have been formed to advance the interests of the UAV community. Association for Unmanned Vehicle Systems International; http://www.auvsi.org/Home.

[7] Colin Cahoon, *Low Altitude Airspace: A Property Rights No-Man's Land*, 56 J. AIR L. & COM. 157, 161 (1990).

[8] *Id*; *see also* R. WRIGHT, THE LAW OF AIRSPACE 11-65 (1968).

[9] Air Commerce Act of 1926, P.L. 69-254, 44 Stat. 568.

[10] Civil Aeronautics Act of 1938, P.L. 75-706, 52 Stat. 973.

[11] Codified as amended at 49 U. S.C. § 40103 (2012).

[12] Codified as amended at 49 U. S.C. § 40101 (2012).

[13] P.L. 85-726; 72 Stat. 737 (1958).

[14] H. Rept. 2360, 85th Cong., 2d Sess. (1958).

[15] S. Rept. 1811, 85th Cong., 2d Sess. (1958).

[16] P.L. 112-95, §332(a)(1).

[17] *Id.* at §332(a)(3).

[18] 14 C.F.R. §91.119(a).

[19] *Id.* at §91.119(b).

[20] *Id.* at § 91.119(c).

[21] *Id.*

[22] 49 U. S.C. § 40102(32).

[23] 14 C.F.R. § 91.119(d).

[24] *See* People v. Sabo, 185 Cal. App. 3d 845, 852 (1986) ("While helicopters may be operated at less than minimum altitudes so long as no hazard results, it does not follow that such operation is conducted within navigable airspace. The plain meaning of the statutes defining navigable airspace as that airspace above specified altitudes compels the conclusion that helicopters operated below the minimum are not in navigable airspace. The helicopter hovering above the surface of the land in such fashion as not to constitute a hazard to persons or property is, however, lawfully operated.").

[25] *See id.* at § 332(b).

[26] FAA, "Unmanned Aircraft Operations in the National Airspace System," 72 Fed. Reg. 6689 (Feb. 13, 2007).

[27] *See id.*

[28] *Id.*

[29] *See generally* FAA "Unmanned Aircraft Systems," *available at* http://www.faa.gov/about /initiatives/uas/cert/.

[30] *See* P.L. 112-95, § 334(a) (instructing the issuance of "guidance regarding the operation of public unmanned aircraft systems to ... expedite the issuance of a certificate of authorization process ... "); *see also* "Certificates of Authorization or Waiver (COA)," *available at* http://www.faa.gov/about/office_org/headquarters_offices/ato/service_units/ systemops/aaim/organizations/uas/coa/.

[31] "FAA makes progress with UAS integration," *available at* http://www.faa.gov/news/updates /?newsId=68004.

[32] 72 Fed. Reg. 6689; *see* 14 C.F.R. §§ 21.191, 21.193 (experimental certificates generally); 14 C.F.R. § 91.319 (operating limitations on experimental certificate aircraft).

[33] P.L. 112-95, § 332(2).

[34] *See* 72 Fed. Reg. 6689; Advisory Circular 91-57, "Model Aircraft Operating Standards" (June 1981), *available at* http://www.faa.gov/documentLibrary/media/Advisory_Circular/91-57.pdf.

[35] P.L. 112-95, § 332(a)(1).

[36] *Id.* at § 332(a)(4).

[37] *Id.* at § 332(a)(2).

[38] *Id.*

[39] *Id.* at § 332(b).

[40] *Id.* at § 332(b)(3).

[41] P.L. 112-95, § 334(a), (c).

[42] *Id.* at § 334(b).

[43] *Id.* at § 336.

[44] *Id.* at § 336(a).

[45] *Id.* at § 336(c).

[46] *Id. at* § 336(a)(5).

[47] *Id.* at § 336(b).

[48] *Id.* at § 332(c).

[49] *Id.* at § 332(c)(2).

[50] Unmanned Aircraft System Test Site Program, 78 Fed. Reg. 12259 (Feb. 22, 2013).

[51] *Id.*

[52] *Id.*; *see* P.L. 112-95, § 332(c)(3).

[53] *See infra* "FAA Regulation of Privacy".

[54] FAA, "UAS Test Site Map," *available at* http://www.faa.gov/about/initiatives/uas/media /UAS_testsite_map.pdf.

[55] United States v. *Causby*, 328 U. S. 256 (1946).

[56] *Id.*

[57] Takings claims filed against state government actors would not be filed under the Fifth Amendment. Rather, they would arise as state constitutional claims. For more information on takings, *see* CRS Report RS20741, *The Constitutional Law of Property Rights "Takings": An Introduction*, by Robert Meltz.

[58] *Causby*, 328 U. S. at 258.

[59] *Id.* at 259.

[60] *Id.* at 260-61.

[61] *Id.* at 260 (citing statutes then codified at 49 U. S.C. §§ 176(a), 403).

[62] *Id.*

[63] *Id.*
[64] *Id.* at 264 (citing Hinman v. Pacific Air Transport, 84 F.2d 755 (9th Cir. 1936)).
[65] *Id.* at 265.
[66] *Id.* at 264.
[67] *Id.* at 263.
[68] *Id.* at 266.
[69] *See e.g.*, Andrews v. United States, 2012 U. S. Claims LEXIS 1644, *10 (explaining that the "The United States Court of Appeals for the Federal Circuit (Federal Circuit) has derived from *Causby* three factors for consideration 'in determining whether noise and other effects from overflights ... constitute a taking.... '"). *But see* Argent v. United States, 124 F.3d 1277, 1284 (Fed. Cir. 1997) (finding a taking claim may be based on "a peculiarly burdensome pattern of activity, including both intrusive and non-intrusive flights").
[70] 49 U. S.C. §40102(32) (2012).
[71] 347 U. S. 590 (1954).
[72] *Id.* at 596.
[73] Griggs v. Allegheny County, 369 U. S. 84, 90 (1962).
[74] *Id.* at 87.
[75] *Id.* at 86-89.
[76] *See, e.g.*, Aaron v. United States, 311 F.2d 798 (Ct. Cl. 1963); Powell v. United States, 1 Cl. Ct. 669 (1983).
[77] Stephens v. United States, 11 Cl. Ct. 352, 362 (1986).
[78] Branning v. United States, 654 F.2d 88, 99 (1981).
[79] *Id.* at 98-99 (citing Palisades Citizens Association, Inc. v. C.A.B, 420 F.2d 188, 192 (D.C. Cir. 1969)).
[80] *See, e.g.*, *Causby*, 328 U. S. 256.
[81] *See, e.g.*, *Griggs*, 369 U. S. 84; Pueblo of Sandia v. Smith, 497 F.2d 1043 (10th Cir. 1974) ("appellant failed to show interference with actual, as distinguished from potential, use of its land.").
[82] *Griggs*, 369 U. S. at 87.
[83] *See, e.g.*, Brown v. United States, 73 F.3d 1100 (1996); *Branning*, 654 F.2d 88.
[84] *Brown*, 73 F.3d 1100.
[85] Stephens v. United States, 11 Cl. Ct. 352 (1986).
[86] Geller v. Brownstone Condominium, 82 Ill. App. 3d 334, 336-37 (1980).
[87] RESTATEMENT (SECOND) OF TORTS §159(2) (1965) (stating that "Flights by aircraft in the airspace above the land of another is a trespass if, but only if, (a) it enters into the immediate reaches of the airspace next to the land, and (b) it interferes substantially with the other's use and enjoyment of the land.").
[88] RESTATEMENT (SECOND) OF TORTS §821D (1979); 2 DAN B DOBBS ET AL., THE LAW OF TORTS §398 (2d ed. 2011).
[89] *See* Jennifer Lynch, *Are Drones Watching You?*, ELECTRONIC FRONTIER FOUNDATION (January 10, 2012), https://www.eff.org/deeplinks/2012/01/drones-are-watching-you; M. Ryan Calo, *The Drone as Privacy Catalyst*, 64 STAN. L. REV. ONLINE 29 (December 12, 2011), http://www.stanfordlawreview.org/sites/default/files/online/articles/64- SLRO-29_1.pdf.
[90] For an analysis of the Fourth Amendment implications of government drone surveillance, see CRS Report R42701, *Drones in Domestic Surveillance Operations: Fourth Amendment Implications and Legislative Responses*, by Richard M. Thompson II.

[91] *See* Press Release, Rep. Ed Markey, Markey Releases Discussion Draft of Drone Privacy and Transparency Legislation (August 1, 2012), *available at* http://markey.house.gov/press-release/markey-releases-discussion-draftdrone-privacy-and-transparency

Drones are already flying in U. S. airspace – with thousands more to come – but with no privacy protections or transparency measures in place. We are entering a brave new world, and just because a company soon will be able to register a drone license shouldn't mean that company can turn it into a cash register by selling consumer information. Currently, there are no privacy protections or guidelines and no way for the public to know who is flying drones, where, and why. The time to implement privacy protections is now.

Id.

[92] Semayne's Case, 5 Co. Rep. 91 (K. B. 1604).

[93] In contesting the use of general warrants by officials of the British Crown, known then as writs of assistance, James Otis argued that "one of the most essential branches of English liberty, is the freedom of one's house. A man's house is his castle; and while he is quiet, he is as well guarded as a prince in his castle." II LEGAL PAPERS OF JOHN ADAMS 142.

[94] *See* State v. Armfield, 9 N.C. 246, 247 (1822).

[95] Note, *The Right to Privacy in Nineteenth Century America*, 94 HARV. L. REV. 1892, 1896 (1981). In an early case from Pennsylvania, in recognizing eavesdropping as an indictable offense, the court noted: "Every man's home is his castle, where no man has a right to intrude for any purpose whatever. No man has a right to pry into your secrecy in your own house." Commonwealth v. Lovett, 4 Pa. L.J. Rpts. (Clark) 226, 226 (Pa. 1831); *see also* State v. Williams, 2 Tenn. 108, 108 (1808) (recognizing eavesdropping as an indictable offense).

[96] Ken Gromley, *One Hundred Years of Privacy*, 1992 WIS. L. REV. 1335, 1351 (1992) (quoting EDWIN EMERY & MICHAEL C. EMERY, THE PRESS AND AMERICA: AN INTERPRETATIVE HISTORY OF THE MASS MEDIA 349-50 (3d ed. 1972).

[97] *Id.* at 1350-51.

[98] William M. Prosser, *Privacy*, 48 Cal. L. Rev. 383, 383 (1960).

[99] Louis D. Brandeis & Samuel D. Warren, *The Right to Privacy*, 4 HARV. L. REV. 193, 205 (1890).

[100] *Id.* at 195.

[101] *Id.* at 205.

[102] *Id.* at 195.

[103] Herbert Spencer Hadley, *Right to Privacy*, 3 N.W. L. REV. 1, 3-4 (1894) ("The writer believes that the right to privacy does not exist; that the arguments in its favor are based on a mistaken understanding of the authorities cited in its support[.]").

[104] *Compare* Roberson v. Rochester Folding Box Co., 171 N.E. 538, 542 (N.Y. 1902) (declining to adopt right of privacy), *with* Pavesich v. New England Life Ins. Co., 50 S.E. 68 (Ga. 1905) (recognizing a right to privacy).

[105] RESTATEMENT (FIRST) OF TORTS §867 (1939).

[106] *See* Louis Nizer, *Right of Privacy – A Half Century's Development*, 39 MICH. L. REV. 526, 529-30 (1940).

[107] Prosser, *supra* note 98, at 385.

[108] RESTATEMENT (SECOND) OF TORTS §§652B (intrusion upon seclusion), 652C (appropriation of name or likeness), 652D (publicity given to private fact), 652E (publicity placing person in false light).

[109] *Id.* at §652B.

[110] Because the use of drones for surveillance primarily concerns the collection, and not necessarily the dissemination, of information, this section will focus on the tort of intrusion upon seclusion, which has no publication requirement for recovery. *Id.* cmt. a.

[111] North Dakota and Wyoming are the only states not to adopt the privacy tort of intrusion upon seclusion. *See* Tigran Palyan, *Common Law Privacy in a Not So Common World: Prospects for the Tort of Intrusion Upon Seclusion in Virtual Worlds*, 38 SW. L. REV. 167, 180 n.106 (2008).

[112] *Id.*

[113] Shorter v. Retail Credit Co., 251 F. Supp. 329, 322 (D.S.C. 1966).

[114] RESTATEMENT (SECOND) OF TORTS §652B (emphasis added).

[115] N.O.C., Inc. v. Schaefer, 484 A.2d 729, 733 (N.J. Super. Ct. Law Div. 1984).

[116] RESTATEMENT (SECOND) OF TORTS §652B cmt. d.

[117] *See, e.g.*, Miller v. National Broadcasting Co., 187 Cal. App. 3d 1463 (Cal. Ct. App. 1986) (videotaping man in his home while being resuscitated after having suffered a heart seizure); Nader v. General Motors Corp., 25 N.Y.2d 560, 570 (1970) (surveilling plaintiff in bank in an "overzealous" manner).

[118] RESTATEMENT (SECOND) OF TORTS §652B.

[119] *See* DOBBS ET AL., *supra* note 88, at §29.

[120] DeAngelo v. Fortney, 515 A.2d 594, 596 (Pa. Sup. 1986); Burns v. Masterbrand Cabinets, Inc., 369 Ill. App. 3d 1006, 1012 (Ill. App. Ct. 2007).

[121] *See, e.g.,* Wolfson v. Lewis, 924 F. Supp. 1413 (E.D. Penn. 1996).

[122] RESTATEMENT (SECOND) OF TORTS §652B cmt. b, illus. 2.

[123] Lovgren v. Citizens First Nat. Bank of Princeton, 534 N.E.2d 987 (Ill. 1989); *see also* Souder v. Pendleton Detectives, 88 So.2d 716, 718 (La. Ct. App. 1956) (peeping into plaintiff's windows); Egan v. Schmock, 93 F. Supp. 2d 1090, 1094-95 (N.D. Cal. 2000) (filming plaintiff and family while in their home).

[124] RESTATEMENT (SECOND) OF TORTS §652B cmt. c.

[125] Prosser, *supra* note 98, at 392.

[126] Schifano v. Green County, 624 So. 2d 178 (Ala. 1993).

[127] Fogel v. Forbes, 500 F. Supp. 1081, 1084, 1087 (E.D. Pa. 1980).

[128] Tellado v. Time-Life, 643 F. Supp. 904, 907 (D.N.J. 1986).

[129] *See* International Union v. Garner, 601 F. Supp. 187, 191-92 (M.D. 1985); Tedeschi v. Reardon v. 5 F. Supp. 2d 40, 46 (D. Mass. 1998).

[130] Jackson v. Playboy Enterprises, Inc., 574 F. Supp. 10, 13 (S.D. Ohio 1983).

[131] McClain v. Boise Cascade Corp., 271 OR 549, 556 (1975). It should be noted that the court also relied on previous case law which held that one who seeks damages for alleged injuries "waives his right to privacy to the extent of a reasonable investigation." *Id.* at 554-555.

[132] *Id.* at 556.

[133] Mojica Escobar v. Roca, 926 F. Supp. 30, 32-33 (D.P.R. 1996).

[134] *Id.* at 35 (citing Dopp v. Fairfax Consultants, Ltd., 771 F. Supp. 494, 497 (D.P.R. 1990)).

[135] GTE Mobilnet of South Texas, LTD. Partnership v. Pascouet, 61 S.W. 3d 599, 605 (Tex. App. 2001).

[136] *Id.* at 618.

[137] *See, e.g.*, Aisenson v. American Broadcasting Co, 220 Cal. App. 3d 146, 162-63 (1990) (holding that broadcast of plaintiff while in his driveway and car was not an intrusion upon seclusion); Wehling v. Columbia Broadcasting System, 721 F.2d 506, 509 (5th Cir. 1983) (holding that broadcast of the outside of plaintiff's home taken from public street was not an

invasion of privacy); Munson v. Milwaukee Bd. of School Directors, 969 F.2d 266, 271 (7th Cir. 1992) (same).

[138] *See* Kramer v. Downey, 684 S.W. 2d 524, 525 (Tex. Ct. App. 1984) ("[W]e now hold that the right to privacy is broad enough to include the right to be free of those willful intrusions into one's personal life at home and at work which occurred in this case.").

[139] RESTATEMENT (SECOND) OF TORTS §652B cmt. c.

[140] Daily Times Democrat v. Graham, 276 Ala. 380, 381 (1964).

[141] *Id.* at 383.

[142] Huskey v. National Broadcasting Co., Inc., 632 F. Supp. 1282, 1285 (1986).

[143] *Id.* at 1286.

[144] *Id.* at 1287-88 (emphasis in original).

[145] Jennifer R. Scharf, *Shooting for the Stars: A Call for Federal Legislation to Protect Celebrities' Privacy Rights*, 3 BUFF. INTELL. PROP. L.J. 164, 183 (2006) ("Modifying intrusion to apply in public places would be necessary in order to provide any relief.").

[146] Abrams v. United States, 250 U. S. 616, 630 (1919) (Holmes, J., dissenting). Justice Stevens described this as a "conflict between interests of the highest order—on the one hand, the interest in the full and free dissemination of information concerning public issues, and, on the other hand, the interest in individual privacy and, more specifically, in fostering private speech." Bartnicki v. Vopper, 532 U. S. 514, 518 (2001).

[147] U. S. CONST. amend. I.

[148] Barry P. McDonald, *The First Amendment and the Free Flow of Information: Towards a Realistic Right to Gather Information in the Information Age*, 65 OHIO ST. L. J. 249, 260 (2004).

[149] Lovell v. City of Griffin, 3030 U. S. 444, 452 (1938).

[150] Watchtower Bible and Tract Soc'y of New York, Inc. v. Vill. of Stratton 536 U. S. 150, 168-69 (2002).

[151] McDonald, *supra* note 148, at 260.

[152] Kleindienst v. Mandel, 408 U. S. 753, 762–63 (1972).

[153] Zemel v. Rusk, 381 U. S. 1, 17 (1965).

[154] *Id.* at 16-17 ("There are few restrictions on action which could not be clothed by ingenious argument in the garb of decreased data flow. For example, the prohibition of unauthorized entry into the White House diminishes the citizen's opportunities to gather information he might find relevant to his opinion of the way the country is being run, but that does not make entry into the White House a First Amendment right.").

[155] Branzburg v. Hayes, 408 U. S. 665, 707 (1972).

[156] *Id.* at 681.

[157] *Id.* at 669.

[158] Cohen v. Cowles Media Co., 501 U. S. 663, 666 (1991).

[159] Dietemann v. Time, Inc., 449 F.2d 245 (9th Cir. 1971).

[160] *Id.* at 249.

[161] *Id.*

[162] Galella v. Onassis, 487 F.2d 986, 991-92 (2d Cir. 1973).

[163] *Id.* at 996-97 (internal citations omitted).

[164] Desnick v. American Broadcast Corporation, 44 F.3d 1345, 1353 (7th Cir. 1995).

[165] For legislation that would regulate public actors, see Thompson, *supra* note 90.

[166] H.R. 1262, 113th Cong. 1st Sess. (1st Sess. 2013).

[167] H.R. 637, 113th Cong (1st Sess. 2013).

[168] As with the enactment of any federal statute, Congress must act within one of its constitutionally delegated powers when creating a federal privacy tort or a crime based on intrusion of privacy. It would appear that Congress could regulate this area under its Commerce Clause power, U. S. Const. art. I, §8, cl. 3, which it acts under when regulating similar federal airspace issues. *See* Braniff Airways v. Nebraska Bd. of Equalization and Assessment, 347 U. S. 590 (1954); United States v. Helsley, 615 F.2d 784 (9th Cir. 1979).

[169] Pearson v. Dodd, 410 F.2d 701, 704 (D.C. Cir. 1969) (quoting Katz v. United States, 389 U. S. 347, 351 (1967)).

[170] Mojica Escobar v. Roca, 926 F. Supp. 30, 35 (D.P.R. 1996).

[171] Federal law does prohibit certain acts of voyeurism on federal property. Section 1801, Title 18 provides: "Whoever, in the special maritime and territorial jurisdiction of the United States, has the intent to capture an image of a private area of an individual without their consent, and knowingly does so under circumstances in which the individual has a reasonable expectation of privacy, shall be fined under this title or imprisoned not more than one year, or both." 18 U. S.C. §1801(a). As discussed in note 168, *supra*, it appears Congress would have the authority to extend this section to voyeurism committed not only on federal property but that committed from federal airspace.

[172] Timothy J. Hortstmann, *Protecting Traditional Privacy Rights in Brave New Digital World: The Threat Posed by Cellular Phone-Cameras and What States Should Do to Stop It*, 111 PENN. ST. L. REV. 739, 742 (2007).

[173] *See, e.g*, GA. CODE ANN. §16-11-61; MONT. CODE ANN. §45-5-223.

[174] NEV. REV. STAT. §200.603.

[175] N.J. STAT. ANN. §2C:18-3c.

[176] WASH. REV. CODE §9A.44.115; *see also* CAL. PENAL CODE §647; R.I. GEN. LAWS §11-64-2.

[177] California Civil Code §1708.8 provides:

> A person is liable for constructive invasion of privacy when the defendant attempts to capture, in a manner that is offensive to a reasonable person, any type of visual image, sound recording, or other physical impression of the plaintiff engaging in a personal or familial activity under circumstances in which the plaintiff had a reasonable expectation of privacy, through the use of a visual or auditory enhancing device, regardless of whether there is a physical trespass, if this image, sound recording, or other physical impression could not have been achieved without a trespass unless the visual or auditory enhancing device was used.

[178] H.R. 3224, H.R. 2448, 105th Cong., 2d sess. (1998).

[179] *See. e.g.,* Press Release, Association for Unmanned Vehicle Systems International, AUVSI to FAA: Focus on your Mission, Proceed with UAS Integration (Nov. 28, 2012) ("As an industry, we support a continued, civil dialogue on privacy, but any such conversations should take place concurrent with the integration. The selection process for the six test sites are a separate issue and should be treated as such. Meanwhile, the FAA should adhere to its mission and do what it does best – focus on the safety of the U. S. airspace – while other, more appropriate institutions consider privacy issues."), *available at* http://www.auvsi.org /AUVSINews/AssociationNews.

[180] *See* Louisiana Pub. Serv. Comm'n v. FCC, 476 U. S. 355, 374 (1986).

[181] Chevron v. Natural Resources Defense Council, 467 U. S. 837 (1984). It should be noted that there is a disagreement among the circuit courts as to whether *Chevron* deference should be accorded to an agency's interpretation of its own jurisdictional statute. *Compare* Hydro Res., Inc. v. EPA, 608 F.3d 1131, 1445-46 (10th Cir. 2010) (en banc) (applying *Chevron*

deference) *with* N. Ill. Steel Supply Co. v. Sec'y of Labor, 294 F.3d 844, 846-47 (7th Cir. 2002) (applying de novo standard). The Supreme Court has granted a petition for writ of certiorari in *City of Arlington v. FCC*, 133 S. Ct. 524 (2012) and may potentially resolve this circuit split. *See* CRS Report WSLG373, *Can an Agency Determine the Scope of its Jurisdiction? Supreme Court Hears Argument Regarding Chevron Deference*, by Daniel T. Shedd.

[182] *Chevron*, 467 U. S. at 842.

[183] *Id.* at 843.

[184] *Id.*

[185] § 332 (b)(1).

[186] § 332(a)(1).

[187] § 332(b)(2).

[188] § 332(b)(1).

[189] *Chevron*, 467 U. S. at 843 (quoting Morton v. Ruiz, 415 U. S. 199, 231 (1974)).

[190] The policy section to FAA's authorizing statue provides that when implementing agency regulations, the FAA shall consider, among other things, the following:

(1) assigning, maintaining, and enhancing safety and security as the highest priorities in air commerce.

(2) regulating air commerce in a way that best promotes safety and fulfills national defense requirements.

(3) encouraging and developing civil aeronautics, including new aviation technology.

(4) controlling the use of the navigable airspace and regulating civil and military operations in that airspace in the interest of the safety and efficiency of both of those operations.

(5) consolidating research and development for air navigation facilities and the installation and operation of those facilities.

(6) developing and operating a common system of air traffic control and navigation for military and civil aircraft.

(7) providing assistance to law enforcement agencies in the enforcement of laws related to regulation of controlled substances, to the extent consistent with aviation safety.

49 U. S.C. § 40101(d).

[191] The "comprehensive plan" must contain, "at a minimum," recommendations on:

(A) the rulemaking to be conducted under subsection (b), with specific recommendations on how the rulemaking will—

(i) define the acceptable standards for operation and certification of civil unmanned aircraft systems;

(ii) ensure that any civil unmanned aircraft system includes a sense and avoid capability; and

(iii) establish standards and requirements for the operator and pilot of a civil unmanned aircraft system, including standards and requirements for registration and licensing;

(B) the best methods to enhance the technologies and subsystems necessary to achieve the safe and routine operation of civil unmanned aircraft systems in the national airspace system;

(C) a phased-in approach to the integration of civil unmanned aircraft systems into the national airspace system;

(D) a timeline for the phased-in approach described under subparagraph (C);

(E) creation of a safe

(F) airspace designation for cooperative manned and unmanned flight operations in the national airspace system;
(G) establishment of a process to develop certification, flight standards, and air traffic requirements for civil unmanned aircraft systems at test ranges where such systems are subject to testing;
(H) the best methods to ensure the safe operation of civil unmanned aircraft systems and public unmanned aircraft systems simultaneously in the national airspace system; and
(I) incorporation of the plan into the annual NextGen Implementation Plan document (or any successor document) of the Federal Aviation Administration.
§ 332(a).

[192] § 332(a)(2).

[193] P.L. 112-95, § 332(c)(1), 126 Stat. 11, 74.

[194] Unmanned Aircraft System Test Site Program, 78 Fed. Reg. 12259 (Feb. 22, 2013).

[195] The FAA has proposed that the OTA include the following privacy requirements:

(1) The Site Operator must ensure that there are privacy policies governing all activities conducted under the OTA, including the operation and relevant activities of the UASs authorized by the Site Operator. Such privacy policies must be available publically, and the Site Operator must have a mechanism to receive and consider comments on its privacy policies. In addition, these policies should be informed by Fair Information Practice Principles. The privacy policies should be updated as necessary to remain operationally current and effective. The Site Operator must ensure the requirements of this paragraph are applied to all operations conducted under the OTA.
(2) The Site Operator and its team members are required to operate in accordance with Federal, state, and other laws regarding the protection of an individual's right to privacy. Should criminal or civil charges be filed by the U. S. Department of Justice or a state's law enforcement authority over a potential violation of such laws, the FAA may take appropriate action, including suspending or modifying the relevant operational authority (e.g., Certificate of Operation, or OTA), until the proceedings are completed. If the proceedings demonstrate the operation was in violation of the law, the FAA may terminate the relevant operational authority.
(3) If over the lifetime of this Agreement, any legislation or regulation, which may have an impact on UAS or to the privacy interests of entities affected by any operation of any UAS operating at the Test Site, is enacted or otherwise effectuated, such legislation or regulation will be applicable to the OTA and the FAA may update or amend the OTA to reflect these changes.
(4) Transmission of data from the Site Operator to the FAA or its designee must only include those data listed in Appendix B to the OTA.

78 Fed. Reg. 12260. Appendix B to the OTA is available at https://faaco.faa.gov/index.cfm /attachment 29581.

The FAA notes that these rules are not permanent but are intended to:

> help inform the dialogue among policymakers, privacy advocates, and the industry regarding broader questions concerning the use of UAS technologies. The privacy requirements proposed here are not intended to pre-determine the long-term policy and regulatory framework under which commercial UASs would operate. Rather, they aim to assure maximum transparency of privacy policies associated with UAS test site operations in order to engage stakeholders in discussion about which privacy issues are raised by UAS operations and how law, public policy, and the industry practices should respond to those issues in the long run.

78 Fed. Reg. 12260.

[196] 78 Fed. Reg. 12260.

[197] Christensen v. Harris County, 529 U. S. 576, 587 (2000).

[198] *Id.*

[199] Skidmore v. Swift & Co., 323 U. S. 134 (1944). In *Skidmore*, the Court was required to determine what level of deference should be accorded the Department of Labor in its issuance of bulletins interpreting a wage provision in the Fair Labor Standard Act. *Id.* at 138. The Court ruled:

> We consider that the rulings, interpretations and opinions of the Administrator under this Act, while not controlling upon the courts by reason of their authority, do constitute a body of experience and informed judgment to which courts and litigants may properly resort for guidance. The weight of such a judgment in a particular case will depend upon the thoroughness evident in its consideration, the validity of its reasoning, its consistency with earlier and later pronouncements, and all those factors which give it power to persuade, if lacking power to control.

Id. at 140.

[200] *Christensen*, 529 U. S. at 588 (citing *Skidmore*, 323 U. S. at 140).

[201] United States v. Mead, 533 U. S. 218 (2001).

[202] *Mead*, 533 U. S. at 230 ("The overwhelming number of our cases applying Chevron deference have reviewed the fruits of notice-and-comment rulemaking or formal adjudication."); *Christensen*, 529 U. S. at 587 ("Here, however, we confront an interpretation contained in an opinion letter, not one arrived at after, for example, a formal adjudication or notice-and-comment rulemaking.") *See also Mead*, 533 U. S. at 231 ("The authorization for classification rulings, and Custom's practice in making them, present a case far removed ... from notice-and-comment process.... ").

[203] The FAA Reform Act provides that in setting up the test sites, the Administrator shall:

(A) safely designate airspace for integrated manned and unmanned flight operations in the national airspace system;

(B) develop certification standards and air traffic requirements for unmanned flight operations at test ranges;

(C) coordinate with and leverage the resources of the National Aeronautics and Space Administration and the Department of Defense;

(D) address both civil and public unmanned aircraft systems;

(E) ensure that the program is coordinated with the Next Generation Air Transportation System; and

(F) provide verification of the safety of unmanned aircraft systems and related navigation procedures before integration into the national airspace system.203The second is FAA's mandate to come up with a comprehensive plan to integrate drones in the national airspace and subsequent rule-making based on this plan.

§ 332(c)(1).

[204] U. S. CONST. art. VI, cl 2.

[205] *See, e.g.*, Crosby v. Nat'l Foreign Trade Council, 530 U. S. 363, 373 (2000); English v. Gen. Elec. Co., 496 U. S. 72, 78-79 (1990); Schneidewind v. ANR Pipeline Co., 485 U. S. 293, 300 (1988).

[206] *See, e.g.*, City of Burbank v. Lockheed Air Terminal, Inc., 411 U. S. 624 (1973); Abdullah v. American Airlines, Inc., 181 F.3d 363 (3d Cir. 1999); San Diego Unified Port Dist. v. Gianturco, 651 F.2d 1306, 1316 (9th Cir. 1981); Price v. Charter Township, 909 F. Supp. 498 (E.D. Mich. 1995).

[207] P.L. 85-726; 72 Stat. 737 (1958).

[208] H.R. Rept. No. 2360, 85th Cong. (1958).

[209] *City of Burbank*, 411 U. S. at 639.

[210] *Id.*

[211] Braniff Airways v. Nebraska Board, 347 U. S. 590 (1954). Additionally, several courts have determined that state law tort claims based on injuries caused by aircraft are not federally preempted. *See, e.g.*, Bieneman v. City of Chicago, 864 F.2d 463 (7th Cir. 1988) (overturning Luedtke v. County of Milwaukee, 521 F.2d 387 (7th Cir. 1975), which ruled that *City of Burbank* preempted application of state tort laws, such as negligence and nuisance, to flights that complied with federal laws and regulations); Greater Westchester Homeowners Association v. City of Los Angeles, 603 P.2d 1329 (Sup. Ct. Cali. 1979).

[212] *See* CRS Report WSLG447, *Congress and the States Grapple with Drones in U. S. Skies*, by Alissa M. Dolan.

[213] "An Act to place a moratorium on the use of unmanned aircraft systems," HB2012, Virginia General Assembly, *available at* http://lis.virginia.gov/cgi-bin/legp604.exe?131+ful+HB 2012ER+pdf.

[214] *See, e.g.*, S. 395, South Carolina General Assembly, 120th Session; S. 524, 77th Oregon Legislative Assembly, 2013 Regular Session; SB 92, Florida Legislature, 2013 Regular Session.

[215] H.B. 912, Texas Legislature, 83rd Session, *available at* http://www.capitol.state LegSess=83R&Bill=HB912.

[216] *Schneidewind*, 485 U. S. at 300.

[217] RESTATEMENT (SECOND) OF TORTS §260.

[218] 18 U. S.C. §2511(1)(a).

[219] 18 U. S.C. §2510(2).

[220] *See, e.g.*, Electromax International, Inc., http://www.electromax.com/penmics.html.

[221] *See* P.L. 112-95, §332(a) (requiring development of a plan within 270 days of enactment of the act, falling in November 2012).

In: Domestic Drones
Editor: Donovan Ambrosio

ISBN: 978-1-63321-456-9

Chapter 2

DRONES IN DOMESTIC SURVEILLANCE OPERATIONS: FOURTH AMENDMENT IMPLICATIONS AND LEGISLATIVE RESPONSES*

Richard M. Thompson II

SUMMARY

The prospect of drone use inside the United States raises far-reaching issues concerning the extent of government surveillance authority, the value of privacy in the digital age, and the role of Congress in reconciling these issues.

Drones, or unmanned aerial vehicles (UAVs), are aircraft that can fly without an onboard human operator. An unmanned aircraft system (UAS) is the entire system, including the aircraft, digital network, and personnel on the ground. Drones can fly either by remote control or on a predetermined flight path; can be as small as an insect and as large as a traditional jet; can be produced more cheaply than traditional aircraft; and can keep operators out of harm's way. These unmanned aircraft are most commonly known for their operations overseas in tracking down and killing suspected members of Al Qaeda and related organizations. In addition to these missions abroad, drones are being considered for use in

* This is an edited, reformatted and augmented version of a Congressional Research Service publication, No. R42701, dated April 3, 2013.

domestic surveillance operations to protect the homeland, assist in crime fighting, disaster relief, immigration control, and environmental monitoring.

Although relatively few drones are currently flown over U. S. soil, the Federal Aviation Administration (FAA) predicts that 30,000 drones will fill the nation's skies in less than 20 years. Congress has played a large role in this expansion. In February 2012, Congress enacted the FAA Modernization and Reform Act (P.L. 112-95), which calls for the FAA to accelerate the integration of unmanned aircraft into the national airspace system by 2015. However, some Members of Congress and the public fear there are insufficient safeguards in place to ensure that drones are not used to spy on American citizens and unduly infringe upon their fundamental privacy. These observers caution that the FAA is primarily charged with ensuring air traffic safety, and is not adequately prepared to handle the issues of privacy and civil liberties raised by drone use.

This report assesses the use of drones under the Fourth Amendment right to be free from unreasonable searches and seizures. The touchstone of the Fourth Amendment is reasonableness. A reviewing court's determination of the reasonableness of a drone search would likely be informed by location of the search, the sophistication of the technology used, and society's conception of privacy in an age of rapid technological advancement. While individuals can expect substantial protections against warrantless government intrusions into their homes, the Fourth Amendment offers less robust restrictions upon government surveillance occurring in public places including areas immediately outside the home, such as in driveways or backyards. Concomitantly, as technology advances, the contours of what is reasonable under the Fourth Amendment may adjust as people's expectations of privacy evolve.

In the 113th Congress, several measures have been introduced that would restrict the use of drones at home. Several of the bills would require law enforcement to obtain a warrant before using drones for domestic surveillance, subject to several exceptions. Others would establish a regime under which the drone user must file a data collection statement stating when, where, how the drone will be used and how the user will minimize the collection of information protected by the legislation.

Introduction

The prospect of drone use in domestic surveillance operations has engendered considerable debate among Americans of various political ideologies.[1]

Opponents of drone surveillance have complained that the use of unmanned aircraft on American soil infringes upon fundamental privacy interests and the ability to freely associate with others.[2] Some are specifically concerned about the possibility of turning military technology inward to surveil American citizens.[3] Proponents have responded by emphasizing their potential benefits, which may include protecting public safety, patrolling our nation's borders, and investigating and enforcing environmental and criminal law violations.[4]

The tension between security and privacy interests is not new, but has been heightened by the explosion of surveillance technology in recent decades. Police officers who were once relegated to naked eye observations may soon have, or in some cases already possess, the capability to see through walls or track an individual's movements from the sky.[5] One might question, then: What is the proper balance between the necessity of the government to keep people safe and the privacy needs of individuals? As some polls suggest, while the public supports drone usage in certain circumstances, they are less enthusiastic about using them as part of routine law enforcement activity.[6]

The Fourth Amendment to the United States Constitution safeguards Americans' privacy and prevents excessive government intrusion by prohibiting "unreasonable searches and seizures."[7] Courts have long grappled with how to apply the text of this 18th century provision to 20th century technologies.

Although the Supreme Court has the final say in the interpretation of the Fourth Amendment and other constitutional safeguards,[8] Congress and, in many cases, the President are free to institute more stringent restrictions upon government surveillance operations.[9]

This report first explores the potential uses of drones in the domestic sphere by federal, state, and local governments. It then surveys current Fourth Amendment jurisprudence, including cases surrounding privacy in the home, privacy in public spaces, location tracking, manned aerial surveillance, and those involving the national border.

Next, it considers how existing jurisprudence may inform current and proposed drone uses. It then describes the various legislative measures introduced in the 113th Congress to address the legal and policy issues surrounding drones. Finally, it briefly identifies several alternative approaches that may constrain the potential scope of drone surveillance.

Background, Uses, and Drone Technology

Drones, also known as unmanned aerial vehicles (UAVs), are aircraft that do not carry a human operator and are capable of flight under remote control or autonomous programming.[10] An unmanned aircraft system (UAS) is the entire system, including the aircraft, digital network, and personnel on the ground.[11] Drones can range from the size of an insect—sometimes called nano drones or micro UAVs—to the size of a traditional jet.[12]

Drones are perhaps most commonly recognized from their missions abroad, including to target and kill suspected members of Al Qaeda and related groups, but they might be used for a variety of other purposes, including for both commercial and law enforcement activities within the United States. In fact, the FAA predicted that 30,000 unmanned aircraft could be flying in U. S. skies in less than 20 years.[13] One reason for this expansion has been a push by Congress for a faster integration of UAVs into U. S. airspace.[14] Most recently, as part of the FAA Modernization and Reform Act of 2012, Congress mandated that the Federal Aviation Administration (FAA) "develop a comprehensive plan to safely accelerate the integration of civil unmanned aircraft systems into the national airspace system."[15] This plan shall provide for integration of UAVs by September 2015.

Drones have been employed domestically by federal, state, and local governments in a range of circumstances. The Department of Homeland Security (DHS) uses them to police the nation's borders to deter unlawful border crossings by unauthorized aliens, criminals, and terrorists, and to detect and interdict the smuggling of weapons, drugs, and other contraband into the country.[16] Within DHS, Customs and Border Protection's (CBP's) Office of Air and Marine (OAM) has flown missions to support federal and state agencies such as the Federal Bureau of Investigation (FBI), the Department of Defense (DOD), Immigration and Customs Enforcement (ICE), the U. S. Secret Service, and the Texas Rangers.[17] According to a recent disclosure by the FAA, several local police departments, state and private colleges, and small cities and towns have also received FAA Certificates of Authorization (COAs) to fly unmanned aircraft domestically.[18] Recently, a police force in North Dakota conducted the nation's first drone-assisted arrest.[19] DHS, in conjunction with local law enforcement agencies, has been testing drone capabilities in a host of other situations including detecting radiation, monitoring a hostage situation, tracking a gun tossed by a fleeing suspect, firefighting, and finding missing persons.[20]

Currently, drones can be outfitted with high-powered cameras,[21] thermal imaging devices,[22] license plate readers,[23] and laser radar (LADAR).[24] In the near future, law enforcement organizations might seek to outfit drones with facial recognition or soft biometric recognition, which can recognize and track individuals based on attributes such as height, age, gender, and skin color.[25] As explained below, the relative sophistication of drones contrasted with traditional surveillance technology may influence a court's decision whether domestic drone use is lawful under the Fourth Amendment.

FOURTH AMENDMENT "SEARCH" JURISPRUDENCE

The Fourth Amendment's story is one of continuity and change. Core values such as privacy and protection from excessive and arbitrary government intrusion are always within its sweep. A continuing question, though, is how the demands of its protection apply to an ever-changing society in which new and pervasive forms of technology are increasingly common. Although there are numerous rules and exceptions throughout the Supreme Court's Fourth Amendment jurisprudence, this section will explore those most pertinent to domestic drone use.

In short, the Fourth Amendment regulates when, where, and how the government may conduct searches and seizures. The Amendment provides, in relevant part: "The right of the people to be secure in their persons, houses, papers, and effects, against unreasonable searches and seizures, shall not be violated[.]"[26] The Fourth Amendment does not apply to all government acts, but only to those that constitute a search. So when does government monitoring constitute a Fourth Amendment "search" for which a warrant is generally required? Initially, courts' assessment focused on the specific area being investigated. Consider the 1928 case *Olmstead v. United States*.[27] There, the Supreme Court held that police wiretaps of the defendant's home telephone did not constitute a Fourth Amendment search because the police did not trespass onto Olmstead's property to intercept his conversation.[28] The Court's thinking at the time was that if the person's home, tangible property, or papers were not physically invaded, then no search in the constitutional sense occurred. Almost 40 years later, the Court shifted focus from property to privacy interests.[29] In *Katz v. United States*, decided in 1967, the Court held that an FBI agent's use of a bug to listen to the private conversations of Mr. Katz while in a telephone booth violated his Fourth Amendment rights.[30] Although he was in a public telephone booth and there was no physical invasion, the Court noted that what a person "seeks to preserve private, even in an area accessible to the public, may be constitutionally protected."[31] One of the modern Fourth Amendment tests relied upon by courts in assessing whether government monitoring constitutes a "search" derives from Justice Harlan's concurrence in *Katz*. It considers whether the person has a subjective expectation of privacy in the area to be searched and whether society is prepared to deem that expectation reasonable.[32]

Although the Court said in *Katz* that the Fourth Amendment "protects people not places,"[33] Justice Harlan noted that determining what "protection it affords people ... requires reference to a '*place*.'"[34] And as Justice Scalia observed when writing for a majority of the Court in *United States v. Jones*, a Fourth Amendment search occurs, "at a minimum," where "the Government obtains information by physically intruding on a constitutionally protected area."[35] The majority in *Jones* indicated that the reasonable expectation of privacy test was never intended to replace the property-based approach used in earlier cases, but merely augment it.[36] So where do individuals enjoy the most Fourth Amendment protection? The least? Why does the location dictate the level of protection? And how does technology affect society's expectation of privacy?

When analyzing domestic drone use under the Fourth Amendment, a reviewing court may be informed by cases surrounding privacy in the home, privacy in public spaces, location tracking, manned aerial surveillance, those involving the national border, and warrantless searches under the special needs doctrine.

Privacy in the Home

The home has always held a central place in American life, and remains the area accorded the greatest Fourth Amendment protection.[37] The Fourth Amendment protects this zone of privacy by ensuring that "the right of the people to be secure in their ... houses ... against unreasonable search and seizure, shall not be violated[.]"[38] In most instances, the Supreme Court has rigorously adhered to this safeguard.[39] For instance, although police officers may make a warrantless arrest of an individual for a felony offense committed in public,[40] they may not step inside his home without a warrant, barring any recognized exception.[41]

In addition to a physical entry and search of the home, police are likewise prohibited from using certain technology to pierce this zone of privacy. In

Kyllo v. United States, government agents used a thermal-imaging device to determine heat patterns inside the home of Danny Kyllo.[42] The Court began with the presumption that a warrantless search of a home is unreasonable.[43] Ultimately, the Court protected this "realm of guaranteed privacy" by holding that obtaining information about the inside of a home that could not otherwise be obtained except by entering the home, through the use of technology not in "general public use," is a "search" covered by the Fourth Amendment.[44]

The home, however, does not provide an absolute shield against government surveillance. As Justice Harlan emphasized in *Katz*, "[A] man's home is, for most purposes, a place where he expects privacy, but objects, activities, or statements that he exposes to the plain view of outsiders are not 'protected' because no intention to keep them to himself has been exhibited." Indeed, in certain instances, police may use their natural senses to conduct warrantless searches of the inside of a home.[45] To fall under this exception—known as the "plain view" doctrine—the police must be in a lawful vantage point when they conduct the surveillance, and the incriminating nature of the evidence must be readily apparent.[46]

Curtilage and Open Fields

Once an individual moves from the confines of the home, he is entitled to different Fourth Amendment considerations. Depending on a host of varying factors, areas outside of the home may be considered "curtilage" or "open fields."[47] The curtilage is the area immediately surrounding the home—an area the Court has granted similar protections as the inside of the home.[48] To determine if an area is curtilage, a court will look at how close the area is to the home; whether the area is within a fence surrounding the home; how the area is used; and whether the area is protected from observation by passersby.[49] Although an area may be deemed curtilage—as with the home—it is not veiled with unconditional constitutional protection. As the Court noted in one aerial surveillance case:

> The Fourth Amendment protection of the home has never been extended to require law enforcement officers to shield their eyes when passing by a home on public thoroughfares. Nor does the mere fact that an individual has taken measures to restrict some views of his activities preclude an officer's observations from a public vantage point where he has a right to be and which renders the activities clearly visible.[50]

On the other hand, the area outside the curtilage is sometimes considered "open fields," which "do not provide the setting for those intimate activities that the Amendment is intended to shelter from governmental interference or surveillance,"[51] and thus are not given similar Fourth Amendment protections. Differentiating between the two is no easy task. In one case, for example, the Ninth Circuit Court of Appeals determined that the defendant's driveway was not curtilage as he had taken no affirmative steps to block it from observation by passers-by.[52] Furthermore, in the fly-over cases discussed *infra*, the Supreme Court has permitted similar searches in both open fields and the curtilage, to some extent eliminating any constitutional difference between the two.

Manned Aerial Surveillance

In a series of cases that provide the closest analogy to UAVs, the Supreme Court addressed the use of *manned* aircraft to conduct domestic surveillance over residential and industrial areas. In each, the Court held that the fly-over at issue was not a search prohibited by the Fourth Amendment, as the areas surveilled were open to public view.

In *California v. Ciraolo*, police received a tip that an individual was growing marijuana in his backyard next to his suburban home.[54] Because two fences blocked their view of the yard, officers flew a fixed-wing aircraft at an altitude of 1,000 feet over the property to conduct a visual inspection. From this vantage point, the officers readily identified with the naked eye marijuana plants growing in the defendant's yard. The Court held that the defendant's expectation of privacy in the area immediately surrounding his home was not reasonable, since "what a person knowingly exposes to the public ... is not a subject of Fourth Amendment protection."[55] "Any member of the public flying in this airspace who glanced down could have seen everything these officers observed," the Court remarked.[56] Much weight was placed on the fact that the plane was at all times in navigable airspace as defined by federal statute.[57]

Similarly, in *Florida v. Riley*, local police received a tip that an individual was growing marijuana in a greenhouse located 10 to 20 feet away from his mobile home.[58] The officers could not see the contents of the greenhouse from the ground, so they flew a helicopter over the defendant's backyard at an altitude of 400 feet. While overhead, an officer saw marijuana plants through a crack in the greenhouse roof. Because the helicopter, like the plane in *Ciraolo*, was in navigable airspace—where any member of the public could have flown—the Court did not consider this a search for which a warrant was required.[59]

> "What a person knowingly exposes to the public ... is not a subject of Fourth Amendment protection."[53]

In the final case of the series, *Dow Chemical v. United States*, the Court was asked whether a theory of "industrial curtilage" would prevent a government agency from conducting aerial surveillance over a 2,000-acre commercial plant.[60] There, after Dow Chemical Co. refused access to the Environmental Protection Agency (EPA), the EPA hired a commercial aerial photographer to take photos of the facility using a precision aerial mapping camera. Having ruled out the argument that the areas surrounding an industrial complex are entitled to the same protection as similar areas surrounding a home, the Court concluded that photographing the plant from navigable airspace was not a search.[61]

Government Tracking

Like the aerial surveillance cases, individuals have reduced—and in some contexts no—Fourth Amendment protection from government tracking of their travel in public places. This has permitted the government to conduct warrantless tracking of a vehicle's movements while traveling on public streets. However, once people enter a private residence, the tracking must cease. Also, using technology to perform pervasive tracking might not meet Fourth Amendment muster.

Consider, for example, these two government tracking cases, *United States v. Knotts* and *United States v. Karo*.[62] In both cases, the government hid a location monitoring device in an item that was then given to the suspects. In *Knotts*, the police tracked the suspect's movements solely while traveling on public roadways.[63] The Court held that people have no reasonable expectation of privacy in their movements on public streets. As such, no Fourth Amendment search occurred, thus no warrant was required. By contrast, in *Karo*, the police tracked a beeper device on public streets *and* while the beeper was in a private residence—a "location not open to visual surveillance."[64] "Indiscriminate monitoring of property that has been withdrawn from public view," the Court declared, "would present far too serious a threat to privacy interests in the home to escape entirely some sort of Fourth Amendment oversight."[65] The Court held the search unlawful.

Although surveillance in public is generally not considered a search, pervasive tracking may cross the line. Take, for instance, the Supreme Court's recent decision in the GPS tracking case *United States v. Jones*.[66] In that case, the Court held that the attachment and monthlong tracking of a GPS device on an individual's vehicle constituted a trespass, and hence a Fourth Amendment search.[67] The Court grounded its decision in the property-based approach to assessing what constitutes a "search" under the Fourth Amendment, which had been more prevalent in the late 19th and early 20th century cases involving relatively unsophisticated technology.

The Court's focus on whether the attachment of a tracking device constitutes a trespass triggering Fourth Amendment protections is not necessarily applicable to drone surveillance. However, in two separate concurring opinions that together made up five members of the Court, an alternative framework was proposed that may have more far-reaching implications for the domestic use of UAVs.

Justice Alito, concurring in the Court's judgment, and joined by Justices Ginsburg, Breyer, and Kagan, would have held that "the use of longer term GPS monitoring in investigations of most offenses impinges on expectations of privacy. For such offenses, society's expectation has been that law enforcement agents and others would not— and indeed, in the main, simply could not—secretly monitor and catalogue every single movement of an individual's car for a very long period." Justice Sotomayor, joining the controlling opinion in *Jones*, but also concurring separately, noted that although following people for a short period of time conveys little information about them, tracking an individual for an extended period "reflects a wealth of detail about her familial, political, professional, religious, and sexual associations."[69] Thus, the length of time an individual is kept under surveillance and the breadth of data collected through such surveillance may inform a reviewing court whether a particular surveillance practice constitutes a Fourth Amendment search.[70]

> Tracking an individual for an extended period "reflects a wealth of detail about her familial, political, professional, religious, and sexual associations."[68]

Border Searches

Even more so than surveillance of public places generally, law enforcement agencies are granted significant deference to conduct surveillance at or near American borders. The federal government has a significant interest in protecting American borders from crossings by persons attempting to enter unlawfully, drug trafficking, and, perhaps most importantly, the transit of weapons and persons seeking to do harm to American people and infrastructure.

Congress has granted federal law enforcement agencies significant search powers at the border. Section 287 of the Immigration and Nationality Act (INA), codified at 8 U. S.C. Section 1357, authorizes immigration officers to conduct warrantless searches of any vessel within a reasonable distance from the United States border and any vehicle within 25 miles from a border for the "purpose of patrolling the border to prevent the illegal entry of aliens into the United States."[71] Similarly, 19 U. S.C. Section 482 authorizes customs officers to search vehicles and persons on which or whom they have reasonable cause to believe are carrying goods unlawfully into the United States.[72]

The Supreme Court has likewise acknowledged this federal interest in the borders, observing that "[t]he Government's interest in preventing the entry of unwanted persons and effects is at its zenith at the international border."[73] Again, the touchstone in every Fourth Amendment case is whether the search is *reasonable*.[74] The Court observed in *United States v. Montoya De Hernandez* that "the Fourth Amendment balance of reasonableness is qualitatively different at the international border."[75] "Routine searches," the Court continued, "are not subject to any requirement of reasonable suspicion, probable cause, or warrant."[76] "Routine" searches have included pat downs for weapons or contraband,[77] the use of drug sniffing dogs,[78] and the inspection of luggage.[79] By contrast, "non-routine" searches are those that go beyond a limited intrusion, and require the government official to have (at a minimum) "reasonable suspicion" of wrongdoing.[80] Prolonged detentions,[81] strip searches,[82] and body cavity searches[83] have all been considered non-routine searches.

Unlike searches directly at the border, the Court has shown more reticence in granting law enforcement unfettered discretion to conduct searches near, but not directly at, the border. In *Almeida-Sanchez v. United States*, the defendant's vehicle was stopped and searched by U. S. Border Patrol agents 25 miles north of the U. S.-Mexico border.[84] The agents had neither a warrant nor probable cause, nor even reasonable suspicion, to conduct the search. The government argued that the search was permissible under Section 287 of the Immigration and Nationality Act. A federal statute, the Court noted, cannot trump the Constitution. The Court refused to permit this suspicionless search, as it was conducted neither at the border nor at its "functional equivalent."[85]

Warrants, Suspicionless Searches, and Special Needs

The baseline rule in Fourth Amendment cases is that police must obtain a warrant to search an individual or their property in all but a few limited instances.[86] This rule ensures that an independent judicial officer, rather than a police officer in the field, is determining whether there is probable cause to conduct a search or seizure.[87] Over time, however, the Court has loosened this warrant requirement in instances where a strict showing of individualized suspicion of probable cause would hinder the government from addressing health and safety concerns.[88] In two lines of overlapping cases—administrative searches and "special needs" cases—the Court has balanced the individual's privacy interest against the government's interest to determine if a warrant or any individualized suspicion is required.[89] These two theories permit law enforcement to conduct dragnet searches in certain instances.

Suspicionless general searches are ones "in which the government searches or seizes every person, place, or thing in a specific location or involved in a specific activity based only on a showing of a generalized government interest."[90] Suspicionless searches have been conducted at the national border,[91] private businesses,[92] and police roadblocks.[93] In the roadblock cases, the Court has balanced the government's interest against the individual's privacy interest to determine whether police may stop and question drivers without a warrant or any suspicion of criminal wrongdoing. In *Michigan Dep't of State Police v. Sitz*, the Court upheld the suspicionless stopping and examination of drivers for intoxication at sobriety checkpoints.[94] The Court reasoned that the high incidence of drunk driving balanced against the minimal intrusion on drivers permitted suspicionless checkpoints. In *City of Indianapolis v. Edmund*, however, law enforcement was not permitted to set up a drug interdiction checkpoint.[95] The Court held that searches such as this violated the Fourth Amendment when their "primary purpose" is "to uncover evidence of ordinary criminal wrongdoing."[96] Justice O'Connor observed that if this type of roadblock were allowed "the Fourth Amendment would do little to prevent such intrusions from becoming a routine part of American life."[97]

Application of Fourth Amendment to Drone Surveillance

As evidenced by the foregoing, the constitutionality of domestic drone surveillance may depend upon the context in which such surveillance takes place. Whether a targeted individual is at home, in his backyard, in the public square, or near a national border will play a large role in determining whether he is entitled to privacy. Equally important is the sophistication of the technology used by law enforcement and the duration of the surveillance. Both of these factors will likely inform a reviewing court's reasoning as to whether the government's surveillance constitutes an unreasonable search in violation of the Fourth Amendment.

Location of Search

Based on existing case law, it is reasonable to assume that surveillance of an individual while in his home—an area accorded the greatest Fourth Amendment protection—using technology not in general public use would be an unlawful search absent a search warrant. The Supreme Court in *Kyllo* was particularly concerned about law enforcement's use of powerful equipment to peer inside an individual's home. Currently, UAVs carry high-megapixel cameras and thermal imaging, and will soon have the capacity to see through walls and ceilings.[98] These technologies are not generally available to the public, and under current jurisprudence, their use by law enforcement would probably constitute a search covered by the Fourth Amendment. However, the use of low-powered cameras or other unsophisticated technology to view people and objects in plain view while in their home might not trigger Fourth Amendment protections. The rationale for this notion is that officers are not required to avert their eyes when they see illegal activity in plain view, especially when the subject of the search has taken no affirmative efforts to hide their activity from public view.

Moving beyond the home, it is unclear whether circumstances exist in which the area immediately surrounding the home—for instance, a backyard, a swimming pool, a deck, or a porch—would receive similar protections as the interior of the home if surveilled by drones or other aerial vehicles.[99] Although the Supreme Court has recited on many occasions that a person located in a home's curtilage is accorded similar privacy protections as when inside the home, the aerial surveillance cases arguably constitute an exception to this general principle. In the two aerial cases, *Riley* and *Ciraolo*, the area surveilled was within close proximity of the home, yet the police surveillance at altitudes of 400 and 1,000 feet were not considered a search.

Based on the aerial surveillance cases, it may be reasonable to presume a warrant would *not* be required (nor, perhaps, any suspicion, for that matter) to conduct drone surveillance of most public places for a relatively short period of time. The Supreme Court remarked in *Ciraolo* that the "Fourth Amendment simply does not require the police traveling in the public airways at [1,000 feet] to obtain a warrant to observe what is visible to the naked eye."[100] However, the rarity of drone flights may distinguish their use from surveillance by the piloted aircraft used in the three aerial cases decided by the Court. All three of these cases were premised on the fact that each aircraft was flying in navigable airspace, and that these flights were not "sufficiently rare" to provide a reasonable expectation of privacy in the area to be searched. To this point, Justice White remarked in *Riley* that "there is nothing in the record or before us to suggest that helicopters flying at 400 feet are sufficiently rare in this country to lend substance to respondent's claim that he reasonably anticipated that his greenhouse would not be subject to observation from that altitude."[101] Presently, use of UAVs in U. S. airspace is considerably less common. The FAA has issued only approximately 300 licenses for drone use in U. S. airspace.[102] The general public would likely find it exceedingly unusual for a drone to fly over their homes taking surveillance photographs. This rarity might factor into a reviewing court's determination of whether individuals have a legitimate expectation of privacy from various forms of drone surveillance while in a public place.[103]

The federal government's authority to use unmanned aircraft is undoubtedly at its maximum near U. S. borders. One of the federal government's only affirmative duties is to protect citizens from external harm.[104] This includes securing the borders. The Court has hesitated from interfering with the performance of this duty, and it would in all likelihood demonstrate the same deference when it comes to the use of UAVs. Moreover, the Supreme Court's rulings in border cases have all involved active searches—either a physical search of a vehicle or stopping and questioning a vehicle's passenger. Surveillance by UAVs, on the other hand, may be considered more passive and therefore may be even less likely to run afoul of Fourth Amendment requirements. Drone surveillance does not require any physical manipulation of a person or his things. UAVs also do not require the seizure of a person for any period of time (though drone surveillance may lead to law enforcement physically apprehending a person who is seen engaging in suspected illegal activity). However, the Court has shown some reticence about giving law enforcement carte blanche search power at the border. Roving vehicle patrols and indiscriminate searches in *Almeida-Sanchez v. United States* and *United States v. Brignoni-Ponce* were deemed unconstitutional.[105] It is unclear whether this reticence would extend to drone surveillance along the border if it were to become significantly widespread.

Technology Used

Like location, the technology used by UAVs may be a decisive factor considered by courts in determining whether individuals have a legitimate expectation of privacy in the object or area of the challenged drone search. Technological developments make it increasingly easy to share and acquire personal information about others, oftentimes without their direct knowledge or consent. As surveillance technology advances and becomes ever-present in Americans' lives, people's conception of privacy may tend to oscillate. Justice Scalia, writing for the majority in *Kyllo v. United States*, remarked on this trend:

> It would be foolish to contend that the degree of privacy secured to citizens by the Fourth Amendment has been entirely unaffected by the advance of technology. For example, ... the technology enabling human flight has exposed to public view (and hence, we have said, to official observation) uncovered portions of the house and its curtilage that once were private. The question we confront today is what limits there are upon this power of technology to shrink the realm of guaranteed privacy.[106]

Justice Alito, joined by three other Justices in his *United States v. Jones* concurrence, likewise observed:

> The *Katz* test rests on the assumption that this hypothetical reasonable person has a well-developed and stable set of privacy expectations. But technology can change those expectations. Dramatic technological change may lead to periods in which popular expectations are in flux and may ultimately produce significant changes in popular attitudes. New technology may provide increased convenience or security at the expense of privacy, and many people may find the tradeoff worthwhile. And even if the public does not

> welcome the diminution of privacy that new technology entails, they may eventually reconcile themselves to this development as inevitable.[107]

The crucial question, then, is whether drones have the potential to be significantly more invasive than traditional surveillance technologies such as manned aircraft or low-powered cameras— technologies that have been upheld in previous cases. In this vein, some have asked whether using sophisticated digital platforms on a drone is any different from attaching the same instrument to a lamppost or traditional aircraft.[108] Take, for example, the tracking of license plates. Currently, many states and municipalities employ automatic license plate readers (ALPRs), which are usually mounted on police vehicles or stationary objects along the streets, to take a snapshot of a license plate as a car drives by, and store this information in a large database for possible later use by law enforcement.[109] It is alleged that these devices can be used to track a person's movements when police aggregate the data from a multitude of ALPR stations.[110] A majority of the reviewing federal circuit courts have held that a person has no reasonable expectation of privacy in his license plate number.[111]

However, it appears that no federal court has addressed the constitutionality of the use of ALPRs (whether attached to a drone, manned vehicle, or a stationary device), as opposed to plate numbers collected by a human observer. Nonetheless, the question remains whether attaching an ALPR—or any similar sophisticated technology—to a drone would alter the constitutionality of its use by law enforcement. Some say yes, arguing that the sophistication of drone technology in and of itself "present[s] a unique threat to privacy."[112] Drones are smaller, can fly longer, and can be built more cheaply than traditional aircraft. For instance, defense firm Lockheed Martin's Stalker—a small, electrically powered drone—can be recharged from the ground using a laser.[113] It now has a flight time of more than 48 hours. As this technology advances, it is reported that some drones could theoretically "stay in the air forever."[114] Unlike a stationary license plate tracker or video camera, drones can lock on a target's every move for days, and possibly weeks and months. This ability to closely monitor an individual's movements with pinpoint accuracy may raise more significant constitutional concerns than some other types of surveillance technology.

Furthermore, the technology and sophistication of drones may mark a considerable departure from the traditional technologies used in the three manned aerial surveillance cases decided by the Supreme Court. First, all three holdings in *Ciraolo*, *Riley*, and *Dow Chemical* were premised on naked-eye searches. Chief Justice Burger remarked in *Dow Chemical*: "It may well be, as the government concedes, that surveillance of private property using highly sophisticated surveillance equipment not generally available to the public, such as satellite technology, might be constitutionally proscribed absent a warrant."[115] As noted above, the sophistication of surveillance technology available to drones, such as facial recognition or laser radar which can "see" through walls, may lead some to question the relevance of prior Fourth Amendment jurisprudence concerning more rudimentary forms of surveillance technology.

The sophistication of drones also has the ability to break down any practical privacy safeguard. In the pre-computer age, the greatest privacy protections were neither constitutional nor statutory, but practical. Putting officers everywhere in the community cost too much for local police departments. This acted as a natural barrier to excessive police presence. Drones are not hindered by similar limitations. This is similar to the expansion of GPS technology observed by the five concurring Justices in *United States v. Jones*. There, Justice Sotomayor, writing for herself, noted that because GPS technology was cheaper and performed in a surreptitious manner, it "evades the ordinary checks that constrain abusive law enforcement practices: limited police resources and community hostility."[116] Justice Alito sounded a similar note in *Jones*: "Traditional surveillance for any extended period of time was difficult and costly and therefore rarely undertaken. [GPS devices] make long-term monitoring relatively easy and cheap."[117] Instead of putting more officers in the field, a police force could put drones in the sky for potentially less expense. Already, drones with video capabilities can be purchased from private vendors for a few hundred dollars.[118] This access to inexpensive technology may significantly reduce budgetary concerns that once checked the government from widespread surveillance.

The duration and pervasiveness of drone surveillance—two factors closely associated with the technology employed by law enforcement—may also influence a court's Fourth Amendment analysis. Consider the Fifth Circuit Court of Appeals' ruling in *United States v. CuevasSanchez*.[119] In that case, federal law enforcement agents suspected the defendant was using his home as a drop house for drug traffickers. After obtaining a court order, the agents installed a video camera on a utility pole overlooking the defendant's 10-foot high fence surrounding his back yard. The officers observed the removal of drugs from the gas tanks of several cars parked in the defendant's yard. Based on these observations, the defendant was arrested whereupon the police seized a large amount of marijuana. At trial, the defendant moved to suppress the evidence on the basis that the warrant was defective. However, the court first addressed whether the video surveillance was a search under the Fourth Amendment. In determining that the video surveillance was a search, the panel noted that this "was not a one-time overhead flight or a glance over the fence by a passer-by.... It does not follow that *Ciraolo* authorizes any type of surveillance whatever just because one type of minimally-intrusive aerial observation is possible."[120] Drones have the capability to stay in the air for long periods of time and can hover in one location. Similar to the mounted camera in *Cuevas-Sanchez*, this permits law enforcement to employ drones in prolonged surveillance operations. This capability may sway a court's determination of whether certain types of warrantless drone surveillance are compatible with the Fourth Amendment.

Warrant Requirement and Suspicionless Drone Searches

Applying the Fourth Amendment to drones requires application of the threshold question: was there a search? Again, this will depend on all the factors discussed above—the area of the search, the technology used, and whether society would respect the target's expectation of privacy in the place searched. If a reviewing court concludes that the drone surveillance was not a search, neither a warrant nor any degree of individualized suspicion would be required. If, however, the court concluded there was a search, then a court would ask whether a warrant is required, if one of the exceptions apply, and what level of suspicion, if any, is necessary to uphold the search.

Unless a meaningful distinction can be made between drone surveillance and more traditional forms of government tracking, existing jurisprudence suggests that a reviewing court would likely uphold drone surveillance conducted with no individualized suspicion when conducted for purposes other than strict law enforcement. The Supreme Court has hesitated from interfering in what they see as the executive's function in protecting the health and safety of the American population. As Chief Justice Rehnquist noted in the *Sitz*, the Court does not want

> to transfer from politically accountable officials to the courts the decision as to which among reasonable alternative law enforcement techniques should be employed to deal with serious public danger.... [F]or purposes of Fourth Amendment analysis, the choice among

> such reasonable alternatives remains with the governmental officials who have a unique understanding of, and a responsibility for, limited public resources, including a finite number of police officers.[121]

The Court may defer to law enforcement officials in the drone context also. There are countless instances where the government may seek to utilize drones for health and safety purposes that go beyond mere law enforcement. These may include firefighting, search and rescue missions, traffic safety enforcement, or environmental protection. If, on other hand, surveillance is conducted primarily to enforce the law, a warrant may be required, unless one of the exceptions to the warrant requirement applies.

LEGISLATIVE PROPOSALS IN THE 113TH CONGRESS TO CONSTRAIN DOMESTIC USE OF DRONES

Although the Supreme Court is the final arbiter of the Constitution, Congress and the President can provide for greater regulation of drones than the Fourth Amendment requires. Congress has taken such steps over the years to address government surveillance of communications in transit (commonly known as wiretapping),[122] communications in storage such as e-mails,[123] bank records,[124] and health records,[125] among a host of other private information. Several measures have been introduced in the 113th Congress that would restrict the domestic use of drones, and establish arguably greater constraints on their usage than the Fourth Amendment requires. Several bills were prompted by a general concern for potential privacy intrusions by federal and state law enforcement and executive agencies.[126]

Preserving Freedom from Unwarranted Surveillance Act of 2013 (H.R. 972)

In the 113th Congress, Representative Austin Scott has introduced the Preserving Freedom from Unwarranted Surveillance Act of 2013 (H.R. 972).[127] This bill would require any entity acting under the authority of the federal government to obtain a warrant based upon probable cause before conducting drone surveillance to investigate violations of criminal law or regulations. There are, however, several exceptions to this warrant requirement: (1) to prevent or deter illegal entry of any persons or illegal substances into the United States; (2) when a law enforcement officer possesses reasonable suspicion that under particular circumstances "swift action to prevent imminent danger to life or serious damage to property, or to forestall the imminent escape of a suspect, or destruction of evidence" is necessary; or (3) when the Secretary of Homeland Security determines credible intelligence indicates a high risk of a terrorist attack by a specific individual or organization. H.R. 972 would create a right to sue for any violation of its prohibitions.

Preserving American Privacy Act of 2013 (H.R. 637)

Representative Ted Poe introduced the Preserving American Privacy Act of 2013 (H.R. 637), which would restrict the domestic use of drones.[128] To begin, the bill regulates surveillance of "covered information," which means "information that is reasonably likely to enable identification of an individual" or "information about an individual's property that is not in plain view." H.R. 637 would create a general prohibition on the use of drones to collect covered information or disclose covered information so collected. It does, however, provide the following exceptions:

- *Warrant.* Law enforcement obtains a court-issued warrant and serves a copy of the warrant on the target of the search within 10 days of the surveillance. However, notice need not be provided if it would jeopardize an ongoing criminal or national security investigation.
- *General Order.* Law enforcement obtains a court-issued order based upon "specific and articulable facts showing a reasonable suspicion of criminal activity and a reasonable probability" that the operation "will provide evidence of such criminal activity." The order may authorize surveillance in a stipulated public area for no more than 48 hours which may be renewed for a total of 30 days. Notice of the operation must be provided to the target no later than 10 days after the operation. Alternatively, notice may be provided not less than 48 hours before the operation in a major publication, on a government website, or with signs posted in the area of the operation.
- *Border searches.* Operation is within 25 miles of national border.
- *Consent.* The targeted individual has provided prior written consent.
- *Emergencies.* Emergency situation involves danger of death or serious physical injury, conspiratorial activities threatening the national

security interest, or conspiratorial activities characteristic of organized crime, where a warrant cannot be obtained with due diligence. Law enforcement must then obtain a warrant within 48 hours of such operation.

Any evidence obtained in violation of this act is not admissible in any trial or adjudicative proceeding. Additionally, H.R. 637 requires any governmental entity applying for a certificate or license to operate a UAS to also file a data collection statement with the Attorney General. The statement must include the following:

- Purpose for which the UAS will be used
- Whether the UAS is capable of collecting covered information
- The length of time the information will be retained
- A point of contact for citizen feedback
- The particular unit of governmental entity responsible for safe and appropriate operation of the UAS
- The rank and title of the individual who may authorize the operation of the UAS
- The applicable data minimization policies barring the collection of covered information unrelated to the investigation of crime and requiring the destruction of covered information that is no longer relevant to the investigation of a crime
- Applicable audit and oversight procedures.

Under H.R. 637, the Attorney General may request that the Secretary of Transportation revoke the license or certificate of any entity that fails to file a data collection statement. Further, H.R. 637 contains a provision permitting administrative discipline against an officer who intentionally violates a provision of this act.

Drone Aircraft Privacy and Transparency Act of 2013 (H.R. 1262)

Representative Ed Markey introduced the Drone Aircraft Privacy and Transparency Act of 2013 (H.R. 1262).[129] This bill would amend the FAA Modernization and Reform Act of 2012 to create a comprehensive scheme to regulate government actors' use of drones, including data collection requirements and enforcement mechanisms.

First, this bill would require the Secretary of Transportation, with input from the Secretary of Commerce, the Chairman of the Federal Trade Commission, and the Chief Privacy Officer of the Department of Homeland Security, to study any potential threats to privacy protections posed by the introduction of drones in the national airspace. Next, the bill would prohibit the FAA from issuing a license to operate a drone unless the application for such use included a "data collection statement." This statement would require the following items:

- A list of individuals who would have the authority to operate the drone
- The location in which the drone will be used
- The maximum period it will be used
- Whether the drone would be collecting information about individuals

If the drone will be used to collect personal information, the statement must include the following:

- The circumstances in which such information will be used
- The kinds of information collected and the conclusions drawn from it
- The type of data minimization procedures to be employed
- Whether the information will be sold, and if so, under what circumstances
- How long the information would be stored
- Procedures for destroying irrelevant data.

The statement must also include information about the possible impact on privacy protections posed by the operation under that license and steps to be taken to mitigate this impact. Additionally, the statement must include the contact information of the drone operator; a process for determining what information has been collected about an individual; and a process for challenging the accuracy of such data. Finally, the FAA would be required to post the data collection statement on the Internet.

In addition to the data collection statement, any law enforcement agency which operates a drone must file with the FAA a "data minimization statement." This statement must include policies adopted by the agency that

- minimize the collection of information and data unrelated to the investigation of a crime under a warrant,
- require the destruction of data that is no longer relevant to the investigation of a crime,
- establish procedures for the method of such destruction, and
- establish oversight and audit procedures to ensure the agency operates a UAS in accordance with the data collection statement filed with the FAA.

H.R. 1262 includes several enforcement mechanisms. First, the FAA may revoke a license of a user that does not comply with these requirements. The Federal Trade Commission would have the primary authority to enforce the data collection requirements just stated. Additionally, the Attorney General of each state, or an official or agency of a state, is empowered to file a civil suit if there is reason to believe that the privacy interests of residents of that state have been threatened or adversely affected. H.R. 1262 would also create a private right of action for a person injured by a violation of this legislation.

CONCLUSION

The introduction of drones into American airspace raises many legal and policy questions. For instance, how far can the government go in its attempts to maintain security and ensure that laws are enforced? What level of privacy should Americans expect in an age where technology facilitating the acquisition of personal information expands at a phenomenal pace? Currently, there is a vast body of Fourth Amendment law that governs the circumstances in which law enforcement must obtain a warrant before conducting surveillance.

However, the sheer sophistication of drone technology and the sensors they can carry may remove drones from this traditional Fourth Amendment framework. Beyond the courts and the Constitution, what role should Congress and the President play in regulating the introduction of drones inside the United States? As the integration of drones for domestic surveillance operations quickly accelerates, these questions and others will be posed to the American people and their political leaders.

End Notes

[1] The term "domestic drone surveillance" as used in this report is designed to cover a wide range of government uses including, but not limited to, investigating and deterring criminal or regulatory violations; conducting health and safety inspections; performing search and rescue missions; patrolling the national borders; and conducting environmental investigations.

[2] Letter from Representatives Edward J. Markey and Joe Barton, Co-Chairmen of the Congressional Bi-Partisan Privacy Caucus, to Michael P. Huerta, Acting Administrator of the Federal Aviation Administration (April 19, 2012) ("[I]n addition to benefits, there is also the potential for drone technology to enable invasive and pervasive surveillance without adequate privacy protections."), *available at* http://markey.house.gov/sites/markey.house.gov/files/documents/ 4-19-12.Letter%20FAA%20Drones%20.pdf; AMERICAN CIVIL LIBERTIES UNION, PROTECTING PRIVACY FROM AERIAL SURVEILLANCE: RECOMMENDATIONS FOR GOVERNMENT USE OF DRONE AIRCRAFT 1 (2011), *available at* https://www.aclu.org/files/assets

[3] Mark Brunswick, *Spies in the sky signal new age of surveillance*, STARTRIBUNE (July 22, 2012, 6:26 a.m.), *available at* http://www.startribune.com/local/163304886.html?refer=y.

[4] Some state and local officials have expressed interest in employment of drones for public safety and law enforcement purposes. *See, e.g.*, Brianne Carter, *Gov. Bob McDonnell supports drones policing Virginia*, ABC NEWS (May 30, 2012), *available at* http://www.wjla.com/articles/2012/05/gov-bob-mcdonnell-supports-drones-policing-virginia-76464.html; *Unmanned Aircraft Systems Within the Homeland: Security Game Changer? Hearing Before the Subcomm. on Oversight, Investigations, and Management of the H. Comm. on Homeland Sec.*, 112th Cong. 3 (2012) (statement of William R. McDaniel, Chief Deputy, Montgomery County Sheriff's Office, Conroe, TX) ("UAV systems for public safety agencies are extremely viable, effective, and economical means to enhance the public safety response to critical incidents.").

[5] The Supreme Court remarked in *Kyllo v. United States*, 533 U. S. 27, 37 (2001):

> The ability to 'see' through walls and other opaque barriers is a clear, and scientifically feasible, goal of law enforcement research and development. The National Law Enforcement and Corrections Technology Center, a program within the United States Department of Justice, features on its Internet Website projects that include a 'Radar-Based Through-the-Wall Surveillance System,' 'Handheld Ultrasound Through the Wall Surveillance,' and a 'Radar Flashlight' that 'will enable law officers to detect individuals through interior building walls.'

[6] *U. S. Supports Some Domestic Drone Use, But Public Registers Concern About Own Privacy* 1, Monmouth University Polling Institute (June 12, 2012), *available at* http://www.monmouth.edu.

[7] U. S. CONST. amend IV.

[8] Cooper v. Aaron, 358 U. S. 1, 18 (1958) ("[Marbury v. Madison] declared the basic principle that the federal judiciary is supreme in the exposition of the law of the Constitution, and that principle has ever since been respected by this Court and the Country as a permanent and indispensable feature of our constitutional system.").

[9] In reaction to the Supreme Court's ruling in *United States v. Miller*, 425 U. S. 435 (1976), that the privacy of an individual's bank records were generally not protected by the Fourth Amendment, Congress enacted the Right to Financial Privacy Act, P.L. 95-630, 92 Stat. 3697 (codified at 12 U. S.C. §3401-3422), creating a statutory protection for such records.

[10] DEP'T OF DEFENSE, DICTIONARY OF MILITARY AND ASSOCIATED TERMS 331 (2012). Unless expressly mentioned, the terms "unmanned aerial vehicle," "UAV," "unmanned aircraft system," "UAS," and "drone" are used interchangeably in this report.

[11] *Id.*

[12] See CRS Report R42136, *U. S. Unmanned Aerial Systems*, by Jeremiah Gertler, for a description of the various types of drones currently operated in the United States.

[13] FEDERAL AVIATION ADMINISTRATION, FAA AEROSPACE FORECAST: FISCAL YEARS 2010-2030, at 48 (2010), *available at* http://www.faa.gov/data_research/aviation /aerospace. The FAA has noted that "Federal agencies are planning to increase their use of UAS's. State and local governments envision using UAS's to aid in law enforcement and firefighting. Potential commercial uses are also possible, for example, in real estate photography or pipeline inspection. UAS's could perform some manned aircraft missions with less noise and fewer emissions." *Id.*

[14] *See, e.g.*, Intelligence Reform and Terrorism Prevention Act of 2004, P.L. 108-458, §5102, 118 Stat. 3638, 3732 (requesting that DHS test the feasibility of using unmanned aircraft to patrol the northern border of the United States).

[15] FAA Modernization and Reform Act of 2012, P.L. 112-95, §332, 126 Stat. 11, 73.

[16] *See* CRS Report RS21698, *Homeland Security: Unmanned Aerial Vehicles and Border Surveillance*, by Chad C. Haddal and Jeremiah Gertler.

[17] The OAM mission is to "protect the American people and the Nation's critical infrastructure through the coordinated use of integrated air and marine forces." DEP'T OF HOMELAND SECURITY, OFFICE OF INSPECTOR GENERAL, CBP'S USE OF UNMANNED AIRCRAFT SYSTEMS IN THE NATION'S BORDER SECURITY 2 (2012). These forces are used to "detect, interdict, and prevent acts of terrorism and the unlawful movement of people, illegal drugs, and other contraband toward or across U. S. borders." *Id.*

[18] There are over 300 total, including those issued to the following entities: City of Herrington, KS; Cornell University; Department of Energy Idaho National Laboratory; Eastern Gateway College Community College—Steubenville, OH; Miami-Dade Police Department; Mississippi Department of Marine Resources; North Little Rock Police Department, AR; Ogden Police Department, UT; Ohio University; Seattle Police Department; Texas A&M—Texas Engineering Experiment Station; Texas Department of Public Safety; Texas State University; University of Connecticut; University of Florida; U. S. Department of Agriculture Agricultural Research Service; Utah State University; Virginia Tech. *See* Unmanned Aircraft Systems, Federal Aviation Administration, http://www.faa.gov /about/initiatives/uas/.

[19] *New age of surveillance*, *supra* note 3.

[20] Brian Bennett, *Drones Tested as Tools for Police and Firefighters*, LOS ANGELES TIMES (August 5, 2012 5:00 A.M.), http://www.latimes.com/news/nationworld/nation/la-na-drones-testing

[21] The U. S. Army recently acquired a 1.8 gigapixel camera for use on its drones. This camera offers 900 times the pixels of a 2 megapixel camera found in some cell phones. It can track objects on the ground 65 miles away from an altitude of 20,000 feet. *US Army unveils 1.8 gigapixel camera helicopter drone*, BBC NEWS (December 29, 2011 6:11 p.m.), http://www.bbc.com/news/technology

[22] Infrared cameras, also known as thermal imaging, can see objects through walls based on the relative levels of heat produced by the objects. *See* Draganflyer X6, Thermal Infrared Camera, http://www.draganfly.com/uav-helicopter/ draganflyer-x6/features/flir-camera.php.

[23] This sensor can recognize and permit drones to track vehicles based on license plate numbers. Customs and Border Protection Today, Unmanned Aerial Vehicles Support Border Security (July 2004), http://www.cbp.gov/xp/ CustomsToday/2004/Aug/other/aerial_vehicles.xml.

[24] This sensor produces three-dimensional images, and has the capability to see through trees and foliage. U. S. ARMY, UAS CENTER FOR EXCELLENCE, "EYES OF THE ARMY" US ARMY ROADMAP FOR UNMANNED AIRCRAFT SYSTEMS 2010- 2035, at 83 (2010).

[25] *See* Clay Dillow, *Army Developing Drones that Can Recognize Your Face from a Distance*, POPSCI (September 28, 2011), 5:01 p.m.), *available at* http://www.popsci.com/technology /article/2011-09/army-wants-drones-can-recognizeyour-face-and-read-your-mind.

[26] U. S. CONST. amend IV.

[27] Olmstead v. United States, 277 U. S. 438, 466 (1928).

[28] *Id.*

[29] Katz v. United States, 389 U. S. 347, 351 (1967).

[30] *Id.* at 359.

[31] *Id.* at 351.

[32] *Id.* at 361 (Harlan, J., concurring).

[33] *Id.* at 351.

[34] *Id.* at 361 (Harlan, J., concurring); *see also* Oliver v. United States, 466 U. S. 170 (1984).

> We have frequently acknowledged that privacy interests are not coterminous with property rights. E. g., United States v. Salvucci, 448 U. S. 83, 91 (1980). However, because "property rights reflect society's explicit recognition of a person's authority to act as he wishes in certain areas, [they] should be considered in determining whether an individual's expectations of privacy are reasonable." Rakas v. Illinois, 439 U. S. 128, 153 (1978) (Powell, J., concurring). Indeed, the Court has suggested that, insofar as "[one] of the main rights attaching to property is the right to exclude others, ... one who owns or lawfully possesses or controls property will in all likelihood have a legitimate expectation of privacy by virtue of this right to exclude." *Id.*, at 144, n. 12.

Oliver, 466 U. S. at 189-90.

[35] United States v. Jones, 132 S. Ct. 945, 950 n.3 (2012).

[36] *Id.*

[37] DANIEL SOLOVE, UNDERSTANDING PRIVACY 59 (2008).

[38] U. S. CONST. amend IV.

[39] Silverman v. United States, 365 U. S. 505, 512 (1961) ("At the very core stands the right of a man to retreat into his own home and be free from unreasonable government intrusion.").

[40] United States v. Watson, 423 U. S. 411 (1976).

[41] Payton v. New York, 445 U. S. 573, 603 (1980). This protection not only covers traditional houses, but apartments, Clinton v. Virginia, 377 U.S 158 (1963), and motel rooms, Stoner v. California, 376 U. S. 483 (1964).

[42] Kyllo v. United States, 533 U. S. 27, 29-30 (2001).

[43] *Id.* at 31.

[44] *Id.* at 34.

[45] In one case, police observation of marijuana plants through a crack in the house's siding did not constitute an unlawful search. *United States v. Hammett*, 236 F.3d 1054, 1061 (9th Cir. 2001). In another, the court held that the police officer's observation of contraband through the front dining room window was not unlawful. *United States v. Taylor*, 90 F.3d 903, 909 (4th Cir. 1996).

[46] Coolidge v. New Hampshire, 403 U. S. 443, 466 (1971).

[47] United States v. Hester, 365 U.S 57 (1924) (distinguishing between the doctrines of curtilage and open fields).

[48] The Court has defined curtilage as "the area to which extends the intimate activity associated with the 'sanctity of a man's home and the privacies of life.'" Oliver v. United States, 466 U. S. 170, 180 (1984) (quoting Boyd v. United States, 116 U. S. 616, 630 (1886)).

[49] United States v. Dunn, 480 U. S. 294, 301 (1987).

[50] California v. Ciraolo, 476 U. S. 207, 213 (1986).

[51] Dow Chemical Co. v. United States, 476 U. S. 227, 234-35 (1986) (quoting *Oliver*, 466 U. S. at 179).

[52] Maisano v. Welcher, 940 F.2d 499 (9th Cir. 1991).

[53] *Ciraolo*, 476 U. S. at 213.

[54] *Id.* at 207.

[55] *Id.* at 213.

[56] *Id.* at 213-214. It should be noted that although the police did take photographs with a 35-millimeter camera, the warrant relied on naked-eye observations and not the photographs. Thus, the holding was based on the naked-eye observations, unaided by the camera.

[57] *Id.* at 213 (citing 49 U. S.C. §1304).

[58] Florida v. Riley, 488 U. S. 445, 448 (1989).

[59] *Id.*

[60] *Dow Chemical Co.*, 476 U. S. 227.

[61] *Id.* at 239

[62] United States v. Knotts, 460 U. S. 276 (1983); United States v. Karo, 468 U. S. 705 (1984).

[63] *Knotts*, 460 U. S. at 279. The Court reiterated that its analysis was based on the fact that the beeper was tracked within the vicinity of a home, not actually in the home. *Id*

[64] *Karo*, 468 U. S. at 714.

[65] *Id.* at 715.

[66] United States v. Jones, 132 S. Ct. 945 (2012).

[67] *Id.* at 949. *See generally* CRS Report R42511, *United States v. Jones: GPS Monitoring, Property, and Privacy*, by Richard M. Thompson II.

[68] *Jones*, 132 S. Ct. at 955 (Sotomayor, J., concurring).

[69] *Id.*

[70] Although the two concurring opinions in *Jones* arguably do not constitute binding precedent, lower courts have combined them as a possible alternative holding. *See* United States v. Hanna, No. 11-20678-CR, 2012 WL 279435, at *3 (S.D. Fla. Jan. 3, 2012) (analyzing the issue of Fourth Amendment standing to contest GPS surveillance under both the trespass theory and *Katz*'s privacy test); State v. Zahn, No. 25584, 2012 WL 862707 (S.D. Mar. 14, 2012) (holding that both the trespass approach and the mosaic theory can apply to GPS tracking); *but see* United States v. Bradshaw, No. 1:11-CR-257, 2012 WL 774964 (N.D. Ohio Mar. 8, 2012) (noting that the *Jones* majority did not adopt the mosaic theory).

[71] 8 U. S.C. 1357.

[72] 19 U. S.C. §482. In the 112th Congress, the House of Representatives introduced H.R. 1505, which would enlarge CBP's authority to secure U. S. borders on federal land. H.R. 1505; *see also* John S. Adams, *Border bill would expand Homeland Security powers*, USA TODAY (September 26, 2011).

[73] United States v. Flores-Montano, 541 U. S. 149, 152 (2004).

[74] Cady v. Dombrowski, 413 U. S. 433, 439 (1973).

[75] United States v. Montoya De Hernandez, 473 U. S. 531, 538 (1985).

[76] *Id.*

[77] United States v. Beras, 183 F.3d 22, 24 (1st Cir. 1999) (ruling that pat down of defendant's legs was routine search).

[78] United States v. Kelly, 302 F.3d 291, 294-95 (5th Cir. 2002).

[79] United States v. Okafor, 285 F.3d 842 (9th Cir. 2002).

[80] *Montoya De Hernandez*, 473 U. S. at 541.

[81] *Id.*

[82] United States v. Asbury, 586 F.2d 973, 975 (1978).

[83] United States v. Ogberaha, 771 F.2d 655, 657 (2d Cir. 1985).

[84] Almeida-Sanchez v. United States, 413 U. S. 266, 267-68 (1973).

[85] *Id.* at 272-73 ("For example, searches at an established station near the border, at a point marking the confluence of two or more roads that extend from the border, might be functional equivalents from border searches. For another example, a search of the passengers and cargo of an airplane arriving at a St. Louis airport after a nonstop flight from Mexico would clearly be the functional equivalent of a border search."). Like *Almeida-Sanchez*, Border Patrol agents in *United States v. Brignoni-Ponce* stopped an individual's vehicle as part of a roving patrol solely because the occupants appeared to be of Mexican descent, with no proof of illegal activity. *United States v. Brignoni-Ponce*, 422 U. S. 873, 874-75 (1975). Again, the Court struck down this practice of suspicionless stops, remarking that "[i]n the context of border area stops, the reasonableness requirement of the Fourth Amendment demands something more than the broad and unlimited discretion sought by the Government. Roads near the border carry not only aliens seeking to enter the country illegally, but a large volume of legitimate traffic as well." *Id.* at 882.

[86] Mincey v. Arizona, 437 U. S. 385, 390 (1978) (citation omitted).

[87] Johnson v. Untied States, 333 U. S. 10, 14 (1948) ("The right of officers to thrust themselves into a home is also a grave concern, not only to the individual but to a society which chooses to dwell in reasonable security and freedom from surveillance. When the right of privacy must reasonably yield to the right of search is, as a rule, to be decided by a judicial officer, not by a policeman or government enforcement agent."). Probable cause is found when, looking at the "totality-of-the-circumstances" there "is a fair probability that contraband or evidence of a crime will be found in a particular place." Illinois v. Gates, 462 U. S. 213, 238 (1983).

[88] Eve Brensike Primus, *Disentangling Administrative Searches*, 111 COLUM. L. REV. 254, 256 (2011).

[89] Nat'l Treasury Employees Union v. Von Raab, 489 U. S. 656, 665-66 (1989).

[90] Primus, *supra* note 88, at 263.

[91] United States v. Martinez-Fuerte, 428 U. S. 543, 566 (1976).

[92] United States v. Biswell, 406 U.S 311, 317 (inspection of gun dealer's storeroom) ("We have little difficulty in concluding that where, as here, regulatory inspections further urgent federal interest, and the possibilities of abuse and the threat to privacy are not of impressive dimensions, the inspection may proceed without a warrant where specifically authorized by statute.").

[93] Mich. Dep't of State Police v. Sitz, 496 U. S. 444, 455 (1990).

[94] *Id.* at 455.

[95] City of Indianapolis v. Edmund, 531 U. S. 32 (2000).

[96] *Id.* at 42.

[97] *Id.*

[98] EYES OF THE ARMY, *supra* note 24.

[99] *See generally,* Paul McBride, *Beyond Orwell: The Application of Unmanned Aircraft Systems in Domestic Surveillance Operations*, 74 J. Air. L. & Com. 627, 655-56 ("This implies that although the curtilage does not benefit from the absolute protection afforded to the interior of the home, there is a close relationship between the two, and that technology directed at the home and its curtilage will be subjected to a more skeptical analysis than would be applied in a case involving open fields or industrial areas.").

[100] California v. Ciraolo, 476 U. S. 207, 215 (1986).

[101] Florida v. Riley, 488 U. S. 445, 451-52 (1989).

[102] Federal Aviation Administration, *supra* note 18.

[103] *Ciraolo*, 476 U. S. at 212. However, in determining society's privacy expectations, a reviewing court might also take into consideration the proliferation of aerial mapping such as Google Maps and Google Earth conducted by private actors. *See generally* Google Maps, Street View, http://www.google

[104] U. S. CONST. art. IV, §4 ("The United States shall guarantee to every State in this Union a Republican Form of Government, and shall protect each of them against Invasion[.]").

[105] Almeida-Sanchez v. United States, 413 U. S. 266 (1973); United States v. Brignoni-Ponce, 422 U. S. 873 (1975).

[106] Kyllo v. United States, 533 U. S. 27, 33-34 (2001).

[107] United States v. Jones, 132 S. Ct. 945, 962 (2012) (Alito, J., concurring).

[108] Stanford Law Review Symposium, *Drones—Privacy Paradox: Privacy and its Conflicting Values* (February 2, 2012), http://cyberlaw.stanford.edu/multimedia

[109] *ACLU Seeks Details on Automatic License Plate Readers in Massive Nationwide Request,* AMERICAN CIVIL LIBERTIES UNION (July 31, 2012), http://www.aclu.org/technology-

[110] *Id.*

[111] *See, e.g.*, Olabisiomotosho v. City of Houston, 185 F.3d 521, 529 (5th Cir. 1999) ("A motorist has no privacy interest in her license plate number. Like the area outside the cartilage [sic] of a dwelling, a car's license plate number is constantly open to the plain view of passersby.") (internal citation and quotation marks omitted); United States v. Ellison, 462 F.3d 557, 562 (6th Cir. 2006) ("No argument can be made that a motorist seeks to keep the information on his license plate private. The very purpose of a license plate number, like that of a Vehicle Identification Number, is to provide identifying information to law enforcement officials and others."); United States v. Castaneda, 494 F.3d 1146, 1151 (9th Cir. 2007); United States v. Walraven, 892 F.2d 972, 974 (10th 1989).

[112] *Using Unmanned Aircraft Systems Within the Homeland: Security Game Changer? Hearing Before the Subcomm. on Oversight, Investigations, and Management of the H. Comm. on Homeland Sec.*, 112th Cong. 3 (2012) (statement of Amie Stepanovich, Counsel, Electronic Privacy Information Center).

[113] Mark Brown, *Lockheed uses ground-based laser to recharge drone mid-flight*, WIRED (July 12, 2012), *available at* http://www.wired.co.uk/news/archive/2012-07/12/lockheed-lasers

[114] *Id.* Lockheed has reportedly been working on extending flight times from days to months. *Id.*

[115] *Dow Chemical*, 476 U. S. at 238.

[116] United States v. Jones, 132 S. Ct. 945, 956 (Sotomayor, J., concurring) (internal quotation marks omitted).

[117] *Id.*

[118] *See, e.g.,* Apple Store, Parrot AR.Drone 2.0, http://store.apple.com/us/product/H8859ZM/A /parrot-ar-drone-2-0. This aircraft, is remote-controlled by an iPhone or iPad, and has a high-definition camera that can take both pictures and videos. *Id.*

[119] United States v. Cuevas-Sanchez, 821 F.2d 248 (5th Cir. 1987).

[120] *Id.* at 251.

[121] Mich. Dep't of State Police v. Sitz, 496 U. S. 444, 453-54 (1990).

[122] Omnibus Crime Control and Safe Streets Act of 1968, P.L. 90-351, 87 Stat. 197, 211 (codified at 18 U. S.C. §2510- 2522).

[123] Electronic Communications Privacy Act of 1986, P.L. 99-508, 100 Stat. 1848 (codified at 18 U. S.C. §2701-2712).

[124] Right to Financial Privacy Act, P.L. 95-630, 92 Stat. 3697 (codified at 12 U. S.C. §3401-3422).

[125] Health Insurance Portability and Accountability Act of 1996, P.L. 104-191, 110 Stat. 1936.

[126] Several bills introduced in the 113th Congress aim to restrict the government's use of drones to lethally target U. S. citizens on U. S. soil. See, e.g., No Armed Drones Act of 2013, H.R. 1083, 113th Cong (1st Sess. 2013); S. 505, 113th Cong. (1st Sess. 2013). These bills, however, are beyond the scope of this report.

[127] H.R. 972, 113th Cong. (1st Sess. 2013). Senator Rand Paul filed similar legislation in the 112th Congress, but has not re-introduced it in the 113th Congress. *See* S. 3287, 112th Cong. (2d Sess. 2012). Unlike H.R. 972, S. 3287 included an express exclusionary rule for evidence obtained in violation of the act.

[128] H.R. 637, 113th Cong. (1st Sess. 2013).

[129] H.R. 1262, 113th Cong. (2d Sess. 2013).

In: Domestic Drones
Editor: Donovan Ambrosio
ISBN: 978-1-63321-456-9

Chapter 3

PILOTLESS DRONES: BACKGROUND AND CONSIDERATIONS FOR CONGRESS REGARDING UNMANNED AIRCRAFT OPERATIONS IN THE NATIONAL AIRSPACE SYSTEM*

Bart Elias

SUMMARY

Growing interest in the use of unmanned aerial vehicles (UAVs), particularly for homeland security and law enforcement applications, has spurred considerable debate over how to accommodate these unmanned aircraft and keep them safely separated from other air traffic. Additionally, the use of these pilotless aircraft, popularly referred to as drones, for aerial surveillance and law enforcement purposes has raised specific concerns regarding privacy and Fourth Amendment rights and potential intrusiveness. These issues have come to the forefront in policy debate in response to provisions in the FAA Modernization and Reform Act of 2012 (P.L. 112-95) that require the Federal Aviation Administration (FAA) to begin integrating unmanned aircraft into the national airspace system by the end of FY2015.

---

* This is an edited, reformatted and augmented version of Congressional Research Service, Publication No. R42718, dated September 10, 2012.

While drones have been used extensively by the military and small radio-controlled model aircraft have been around for more than 50 years, advances in more complex vehicle controls and imaging sensor capabilities are spurring public sector and commercial interest in unmanned aircraft for a variety of purposes, including law enforcement, homeland security, aerial imaging, and scientific research. FAA currently approves public entities (such as federal agencies, public universities, and local police departments) to operate UAVs on a case-by-case basis, but growing interest is making this approach increasingly untenable. Moreover, commercial users are seeking authorization to fly drones, but so far FAA has only allowed test and demonstration flights by manufacturers. FAA faces a number of challenges to address anticipated growth in demand for civilian UAV operations and develop regulations governing the certification and operation of unmanned aircraft systems in domestic airspace.

Congress has generally supported efforts to integrate unmanned aircraft into the national airspace system and foster growth in the unmanned aircraft industry. It enacted extensive provisions in P.L. 112-95 that are designed to streamline and accelerate the operation of unmanned aircraft in domestic airspace by both public entities and commercial operators. Notably, that law requires FAA to issue regulations pertaining to the operation of small UAVs (weighing less than 55 pounds) and requires FAA to create and implement a plan to begin the integration of unmanned aircraft into the national airspace system by the end of FY2015. Toward that goal, the act requires FAA to establish six test ranges throughout the United States to study unmanned aircraft integration technical issues.

The act establishes an ambitious timeline for FAA to grapple with and resolve a number of complex issues regarding the safety and security of unmanned aircraft operations. Furthermore, aircraft operators have expressed specific concerns that drone operations should not result in airspace restrictions or other measures that could limit accessibility to the national airspace system.

In addition to these various challenges, the privacy implications and potential intrusiveness of drone operations have emerged as a significant issue before Congress. Civil liberties and privacy groups have cautioned that voluntary industry measures, including a code of conduct to, among other things, respect privacy, are inadequate to assure that drones will not be misused in ways that could infringe upon the privacy of individuals and intrude upon their daily activities. Moreover, FAA's authority over specific uses of civilian unmanned aircraft appears limited so long as safety and national security are not compromised, raising additional concerns that future drone operations could lead to complaints and lawsuits over noise, intrusiveness, and interference with the use and enjoyment of public or private property.

Background

In the early 2000s, military and intelligence use of unmanned aircraft systems (UASs)[1] in Afghanistan and Iraq spurred interest in potential domestic use of these systems. The safe integration of unmanned aircraft in the crowded airspace above the United States poses a number of regulatory challenges for the Federal Aviation Administration (FAA) and raises some unique legal and ethical questions. These issues have come to the forefront following enactment of the FAA Modernization and Reform Act of 2012 (P.L. 112-95), which requires FAA to begin integrating unmanned aircraft into the national airspace system by the end of FY2015.

Military Roots

While much of the policy debate surrounding civilian use of unmanned aircraft is new, unmanned aircraft have been used by the military since the early days of aviation. During World War I, the Navy funded research to develop a prototype flying bomb called the Hewitt-Sperry Automatic Airplane. Building on this concept, the first full-scale powered unmanned aerial vehicles, the Curtiss-Sperry Flying Bomb and the Kettering Aerial Torpedo, nicknamed the "Bug," were developed in 1918 as unpiloted bomb delivery platforms. The "Bug" was designed to fly a pre-set course, then shed its wings delivering an explosive payload that would detonate on impact. It was never used in combat, and further development was halted in the 1920s.[2] As evidenced by these early prototypes, the U. S. military has long had an interest in unmanned aircraft, which evolved into the development of several successful guided missile designs during World War II.

The military has used autonomous target drones and remotely piloted vehicles extensively since the 1950s. These vehicles first served as targets for training combat pilots and as decoys in combat arenas, but their roles evolved with advances in sensor technology, allowing them to take on more advanced aerial reconnaissance roles.

In the 1980s, spurred by Israeli initiatives, the military developed more sophisticated unmanned systems with extensive ground-based command and control capabilities. Unmanned aircraft were used by U. S. forces for intelligence missions in the Balkans in the late 1990s, and in Afghanistan and Iraq for reconnaissance and surveillance as well as, more recently, for combat missions.[3] In support of military research, development, testing, and training, the military now operates unmanned aircraft extensively in U. S. airspace. The military services work closely with FAA to designate special use airspace for conducting unmanned aircraft operations, typically over military bases or sparsely populated land.

From Longtime Hobby to a Burgeoning Civilian Market

Until recently, civilian interest in operating unmanned aircraft in U. S. airspace had largely been limited to long-standing hobbyist use of radio-controlled model aircraft. The Academy of Model Aeronautics, a national organization representing model aviation enthusiasts, was founded in 1936 and claims a membership of more than 150,000.[4] It serves as the chartering organization for more than 2,400 model airplane clubs across the United States. At present, radio-controlled model aircraft operate under a voluntary standard that generally permits flights only below 400 feet above the surface. These guidelines specify that model aircraft be flown away from populated areas and not in close proximity to full-scale aircraft. They further request coordination with air traffic facilities when planning operations within 3 miles of an airport and suggest that model aircraft be adequately flight tested before being operated in front of spectators.[5]

While most radio-controlled model aircraft are powered by small propeller engines and weigh only a few pounds, some are jet-powered scale models weighing over 100 pounds. The distinction between functional model aircraft and small unmanned aerial vehicles is, therefore, largely based on an aircraft's use rather than its size or capabilities: whereas model aircraft are regarded as being used for recreational purposes, UAVs serve public use missions, such as law enforcement and disaster relief, and commercial applications. The prospect of robust growth in use of UAVs had triggered more formal analysis of the process for granting approvals and development of regulations for UAV operations. The challenges associated with safely integrating civilian UAVs into domestic airspace are discussed in detail below (see "Domestic Airspace Integration Issues").

Flight missions considered to be "dirty, dull, or dangerous" are regarded as prime candidates for the use of unmanned aircraft. Examples include aerial surveillance missions for homeland security, border protection, and law enforcement; highway traffic monitoring; forest fire scouting; disaster response; aerial applications of pesticides; pipeline and transmission line inspection; aerial surveying and geospatial imaging; atmospheric and environmental science; wildlife and natural resources management; scientific data collection; and hurricane and severe storm monitoring. Some industry experts foresee eventual use of unmanned aircraft for cargo transport. At this point, however, passenger-carrying UAVs are not on the horizon.

Industry analysts anticipate a robust market for unmanned aircraft systems, although the extent to which civilian sales will contribute to this market is highly dependent on how the regulation of civilian drones proceeds. The Teal Group, an aviation and aerospace consulting firm, predicted that, over the next ten years, annual spending on unmanned aircraft systems and sensor payloads will increase by 73% and worldwide spending on unmanned aircraft systems will total $89 billion, with the United States accounting for 62% of the research and development spending and 55% of procurement totals.[6] Similarly, VisionGain, a UK-based business information provider, foresees strong demand for unmanned aircraft payloads and subsystems, as well for UAV flight training and simulation.[7]

Unmanned Aircraft Designs

Unmanned aircraft come in all shapes and sizes. They range in size from bug-sized "nano drones" to vehicles the size and weight of large business jets. Small UAVs have an upper weight limit of 55 pounds. Typical civilian drones are likely to fall in the range of a few pounds up to that weight limit, while medium-sized unmanned aircraft, like the MQ9 Predator 2/Reaper (with a 65-foot wingspan and a maximum takeoff weight of about 10,000 pounds) may be used on more specialized, medium-altitude and high-altitude, long-duration missions like border and maritime patrols and environmental research missions. Unmanned aircraft include fixed-wing airplanes, helicopters, rotorcraft, and even blimps. Unmanned aircraft may also include certain vehicles, referred to as optionally piloted vehicles, that are capable of being flown by a pilot as a conventional aircraft or being operated remotely or autonomously depending on mission conditions, needs, and requirements. Figure 1 provides a montage of various unmanned systems illustrating a range of different vehicle sizes and designs including (clockwise from top left) the hand-launched AeroVironment RQ-20 Puma; an unmanned version of the Kaman K-MAX external load helicopter operated by the U. S. Marine Corps; the Honeywell RQ-16 "Tarantula" T-Hawk; the AAI RQ-2 Pioneer; and the "Altair" variant of the General Atomics MQ-9 Predator 2/Reaper.

Sources: Clockwise from top left, U. S. Army, U. S. Marine Corps, U. S. Army, U. S. Army, National Oceanographic and Atmospheric Administration (NOAA).

Figure 1. A Variety of Small to Medium-Sized Unmanned Aircraft Designs.

While a substantial portion of this market will be accounted for by the defense sector, manufacturers are increasingly looking toward the civilian sector, fearing that military sales may stagnate as operations in Afghanistan are scaled back and U. S. defense budgets grow tighter. However, some forecasters have concluded that growth in the civilian market is unlikely until a regulatory framework allowing routine UAS operations is established.[8] FAA anticipates that once such regulations are put in place, roughly 10,000 active civilian UASs will commence operations within the first five years.[9]

It is important to distinguish public-use missions from purely civilian activities involving UAVs, although certain commercial operations, such as utility line inspection, may serve a public benefit. The distinction matters because aircraft owned or leased by federal and state agencies, counties, and municipalities, are regulated differently from civil aircraft operated by private individuals and corporations.[10] Although the regulatory distinctions are complex, in general, FAA oversight and authority over public-use aircraft are more limited. For this reason, some public-sector entities, particularly homeland security and law enforcement agencies and public universities, are already making limited use of UAVs while the commercial market awaits FAA regulatory action.

Congressional Interest and Related Legislation

Congress has generally been supportive of efforts to integrate unmanned aircraft into the national airspace system and foster growth in the unmanned aircraft industry. Through oversight and legislation, it has pushed FAA to accelerate its regulatory processes regarding the certification and operation of unmanned aircraft.

UAVs in Vision 100—The Century of Aviation Reauthorization Act

In 2003, Congress included language in Vision 100—The Century of Aviation Reauthorization Act (P.L. 108-176) specifying that the Next Generation Air Transportation System (NextGen), which FAA is developing to handle air traffic control, shall "accommodate a wide range of aircraft operations, including airlines, air taxis, helicopters, general aviation, and *unmanned aerial vehicles*."[11] That language set the stage for ongoing policy and technical consideration of how to best accommodate unmanned aircraft in the future airspace system. While unmanned aircraft were only briefly mentioned in the act, the specific language requiring that they be accommodated in the design of NextGen served as an important step toward recognizing that drones would likely play a significant role in the aviation system of the future. Nearly a decade later, unmanned aircraft have evolved considerably, and the importance of considering them in aviation regulation and the design and operation of the national airspace system has become a major policy and technical consideration.

Provisions of the FAA Modernization and Reform Act of 2012

During debate over the most recent FAA reauthorization, the integration of unmanned aircraft in national airspace was an issue of considerable interest. The resulting legislation, the FAA Modernization and Reform Act of 2012

(P.L. 112-95) contained extensive provisions designed to promote and facilitate the use of civilian unmanned aircraft. These included mandates for

- development of an integration plan that is to commence by the end of FY2015, if not sooner, along with a five-year roadmap for achieving integration objectives;
- selection of six test sites to study UAV integration into the national airspace system;
- designation of certain permanent areas in the Arctic where small unmanned aircraft may operate 24 hours per day for commercial and research purposes, including flights conducted beyond line-of-sight;
- a simplified process for issuing authorizations for entities seeking to operate public UASs in the national airspace system;
- a collaborative process to incrementally expand airspace access as technology matures and safety data and analysis become available, and to facilitate public agency access to UAS test ranges;
- a requirement to develop and implement operational and certification requirements for public UASs by December 31, 2015; and
- an exemption from rules and regulations pertaining to the operation of unmanned aircraft for model aircraft weighing 55 pounds or less that are flown within visual line-of-sight strictly for hobby or recreation.

CURRENT FAA POLICY AND OVERSIGHT

The provisions in P.L. 112-95 reflect a general view that FAA's slow progress to date on addressing regulatory requirements for unmanned aircraft may be a barrier to civilian drone operations. While FAA has been studying unmanned aircraft operations for almost a decade, it continues to address requests to operate unmanned aircraft on a case-by-case basis. This approach is becoming increasingly untenable as growing numbers of public and commercial entities seek authorization to operate unmanned aircraft in domestic airspace.

On February 13, 2007, FAA issued a notice of policy declaring that "no person may operate a UAS in the National Airspace System without specific authority."[12] This policy applies to both public and private unmanned aircraft. FAA currently has two methods for granting authority to operate unmanned aircraft, depending on whether the operator is a public or a private entity.

Current Approvals Process

Unmanned aircraft operated by the military or other federal, state, or local agencies must obtain a Certificate of Waiver or Authorization (COA) from FAA. Since 2006, FAA has issued COAs to more than 100 unique public entities throughout the United States, ranging from local fire departments to research laboratories. Since the COAs have narrow scopes and durations, entities must obtain new authorizations every few months. In 2009, FAA issued 146 such COAs. In 2010, the number of COAs issued grew to 298.[13] Although FAA has streamlined its approval process, as required by P.L. 112-95, growing interest in operating unmanned aircraft from public entities is expected to place strains on the current case-by-case approach to authorization.

Additionally, in much the same manner as manned aircraft developers and builders must obtain FAA certification to test fly their aircraft, a number of drone developers have received special airworthiness certificates in the experimental category. Special airworthiness certificates, which offer the only legal route for private entities to operate unmanned aircraft for commercial purposes, have been issued on a limited basis for flight tests, demonstrations, and training.

FAA has raised concerns that other civilian users are operating commercial UAVs under the voluntary guidelines issued in the early 1980s that were intended to apply only to recreational users of model aircraft. As previously described, these guidelines advise such users to maintain altitudes lower than 400 feet above the ground, select sites away from populated and noise-sensitive areas, give right of way to full-scale aircraft, and advise airport operators and air traffic facilities if operating within 3 miles of an airport. The FAA statement of policy clarifies that these general guidelines alone are not sufficient for commercial operators of unmanned aircraft, regardless of the size of such aircraft. The FAA did, however, indicate that it has undertaken a safety review to determine whether certain small, slow-moving unmanned aircraft could be safely operated under a similar set of guidelines without requiring special airworthiness certification.

Regulating Small Unmanned Aircraft

Developers of small-scale unmanned aircraft are concerned that FAA has been moving too slowly on measures to safely accommodate these types of aircraft. Since 2007, industry has been pushing for separate regulations for "ultralight, low-speed, short-life UASs that will not be flying over populated areas or in controlled airspace," arguing that a one-size-fits-all approach to regulation would not likely be effective given the wide range of systems being considered for civilian use.[14]

Other than streamlining the case-by-case approvals process, FAA has not taken any formal action addressing the regulation of civilian small unmanned aircraft. After several delays, FAA plans to release a proposed rule regarding the certification and operation of small unmanned aircraft systems (sUAS) in October 2012, as a step toward meeting the August 2014 statutory deadline for issuance of a final rule. Additionally, P.L. 112-95 directed FAA to assess whether certain unmanned aircraft operations do not pose a safety hazard or a threat to national security and can therefore safely be allowed. Actions to address this mandate are currently in progress.

Wide Scale Integration of Unmanned Aircraft in Domestic Airspace

UAV manufacturers and users are seeking a regulatory structure for the certification of UAV systems and approval for operation in domestic airspace. Ideally, operators of approved systems want the ability to "file and fly," meaning that they would be granted vehicle certification with broad operating authority akin to current manned aircraft certification standards. Such regulation would potentially allow unmanned aircraft operators to file routine flight plans, or in some cases simply carry out flight operations without any specific notification requirements, much as manned aircraft do.[15] While FAA's approach to address the mandates set forth in P.L. 112-95 regarding unmanned aircraft integration into the national airspace system is yet to be defined, it is most likely that FAA regulation and oversight of UAVs will adopt an evolving, risk-based approach toward this end goal of seamless integration.

As a first step, P.L. 112-95 mandated that FAA identify six test sites to specifically test concepts and technologies for integrated unmanned aircraft operations. FAA held public meetings and webinars and solicited public comments on the selection of test sites in March 2012, but has not yet gone through a formal source selection process or announced further details regarding test site selection, even though it was mandated to identify the sites by the summer of 2012 and have at least one site operational by February 2013.

Under P.L. 112-95, FAA is also required to develop a comprehensive plan to safely accelerate integration of civilian unmanned aircraft into the national airspace system as soon as practical but not later than September 2015. Meeting this mandate will require extensive research and regulatory action over the next three years.

DOMESTIC AIRSPACE INTEGRATION ISSUES

Key challenges to drone integration in domestic airspace include safety concerns, potential security risks, and concerns regarding airspace restrictions and possible disruptions to manned flight operations.

Safety Concerns

Threshold criteria for integrated unmanned aircraft operations are that UAVs do not pose undue risk other aircraft or compromise safety to persons or property on the ground.[16] To meet safety objectives, UASs will require technology and standard procedures for sensing and avoiding other air traffic under all possible scenarios, including lost communications. System reliability and human factors are important systems design considerations, as are training and qualification standards for drone pilots, unmanned aircraft systems operators, and other safety-critical personnel.[17]

Sensing and Avoiding Other Air Traffic

Domestic airspace accommodates more than 70,000 flights per day[18] at a variety of altitudes, including low- and high-altitude military training flights, high-altitude air carrier and business jet flights, medium-altitude commuter and general aviation flights, and low-altitude recreational and sightseeing flights and helicopter operations. The risk of collision between these users and unmanned aircraft must be adequately mitigated before unmanned aircraft can routinely utilize the national airspace system.

The Government Accountability Office (GAO) concluded that no suitable technology is currently available to provide unmanned aircraft, particularly small UAVs, with the detect, sense, and avoid requirements needed to safely operate within the national airspace system. GAO noted that small unmanned aircraft pose a particular challenge because they operate at low altitudes. Many other aircraft operating at these altitudes do not use electronic transponders to broadcast their position and altitude, and in any case many small UAVs lack the ability to receive transponder signals. The needed equipment is simply too large and heavy to install on many small UAVs.[19]

Currently, these limitations substantially restrict UAV operations to line-of-sight scenarios, where operators on the ground or spotters in chase planes can provide the necessary capabilities to detect and avoid other air traffic. FAA is currently evaluating options for routinely allowing small unmanned aircraft to use line-of-sight as an acceptable means to detect and avoid manned aircraft under a regulatory regime for small UAVs.

For more sophisticated medium and large-sized UAVs seeking approval for operations beyond line of sight, technology advancements are needed to assure safety in an environment shared with manned flights. Remote sensing capabilities, including onboard cameras, airborne radars, and equipment to interrogate aircraft transponder signals (similar to the traffic collision avoidance systems on jetliners) can combine to provide operators with robust air traffic information.

An emerging technology, Automated Dependent Surveillance-Broadcast (ADS-B) will be required on manned aircraft by 2020 and will serve as the principal means for aircraft tracking under NextGen. Using ADS-B, aircraft will broadcast precise global positioning system (GPS) location information to air traffic controllers and other air traffic, potentially including unmanned aircraft. While ADS-B may provide unmanned aircraft operators with the capability to gather precise position and flight path information for nearby air traffic in the future, backup systems will likely be needed because ADS-B transmission and reception may not be guaranteed all the time.

FAA has not specifically approved any technology or suite of technologies as being sufficient to provide acceptable "sense and avoid" capabilities.[20] Part of the challenge is that existing technologies do not assure avoidance capabilities under all operational conditions, including autonomous UAV operations or in situations when UAVs lose their command guidance links with ground control facilities. Standardized procedures for responding when UAV guidance has been lost are currently lacking, but will be needed to ensure that air traffic controllers and airspace managers can redirect nearby traffic and mitigate collision risks.

No single technology is likely to address the complex sense and avoid requirements that are critical for unmanned aircraft integration. FAA has been working closely with RTCA, Inc., which functions as a federal advisory committee on aviation technologies, to develop consensus standards regarding minimum aviation system performance standards for sense and avoid technologies as well as for command, control, and communications systems for unmanned aircraft systems.[21] RTCA has been performing this work under its standards committee on unmanned aircraft systems (SC-203), which is scheduled to complete its work on these issues by December 2013 and intends to publish recommended standards soon thereafter. These standards, along with findings from unmanned aircraft integration testing, will likely form the basis of FAA regulations and guidance regarding sense and avoid technologies and UAV operating procedures.

Mitigating Risks to Persons and Property on the Ground

In addition to risks to other air traffic, unmanned aircraft operations may pose a risk to persons and property on the ground. Thus far, unmanned aircraft testing and operational use, such as for military testing and training and for border and maritime patrol operations, have been conducted largely over sparsely populated areas. In the future, however, law enforcement and commercial users are expected to undertake flights over densely populated areas, a prospect which raises specific concerns over safety procedures. For example, to minimize the likelihood of a crash when communications, command, and control linkages between the vehicle and the ground control station are disrupted, UAVs may need the ability to autonomously return to base.

Drone crashes, such as the 2006 crash of an MQ-9 Predator B drone operated by Customs and Border Protection (CBP) near Nogales, NM, and the more recent crash of a Navy RQ-4A Global Hawk near Salisbury, MD in June, 2012, have raised public concern over the safety of unmanned aircraft operations. While drone safety has improved considerably over the past decade, the accident rate for unmanned aircraft is still far above that of manned aircraft, although direct comparisons are difficult, given that most drone operations examined have involved riskier flying during development and testing and in war and conflict zones.[22]

The risk posed to persons and property on the ground is a function of both crash likelihood and the potential consequences in terms of loss of life, injury, or property damage. While smaller UAVs may be expected to crash more frequently, the potential for catastrophic consequences is less given that these vehicles do not weigh enough or carry enough fuel to cause major damage on the ground. On the other hand, larger UAVs, like the MQ-9 Reaper or the RQ-4 Global Hawk, can potentially cause as much damage as a mid- to large-sized corporate jet. However, safety considerations in the design and operation of these more complex systems may reduce the likelihood of a crash. Additional procedures that can be incorporated into safety regulations for unmanned aircraft systems may further mitigate flight risks. These procedures may include

- formal risk assessments for systems certification and mission planning;
- development of ground impact models and mitigation plans to reduce risks to persons and property on the ground;[23] and
- structured training and certification requirements for unmanned aircraft pilots, systems operators, and other safety critical personnel.

Human Factors Considerations

The issue of training and certification requirements for unmanned aircraft systems personnel raises a much broader issue regarding the role of humans in these systems. Indeed, the term *unmanned* aircraft system (UAS) is a misnomer: while the aircraft themselves may be unmanned, the systems needed to operate them safely depend extensively on human interaction. Although drones operate without a pilot on board, human performance is a major consideration in setting policy for the integration of unmanned aircraft systems in domestic airspace. Key human factors to be considered include operator interfaces and controls and the training and qualifications of drone pilots, systems operators, and other safety critical personnel.

The previously mentioned 2006 crash of a UAV operated by Customs and Border Protection (CBP) illustrates the importance of these human factors considerations. On April 25, 2006, a Predator B (MQ-9) drone crashed in a remote area along the U. S.-Mexico border near Nogales, AZ following a loss of engine power. The National Transportation Safety Board (NTSB) determined that the probable cause of the mishap was the drone pilot's failure to follow appropriate procedures when switching to an alternate control console in the ground control station following a computer malfunction.[24] The error resulted in the pilot inadvertently cutting off the vehicle's fuel supply.

CBP had only been operating the Predator B aircraft since September 2005, eight months prior to the mishap. NTSB concluded that during this time CBP was providing a minimal amount of operational oversight of its UAS program and cited its inadequate surveillance of the program as a contributing factor in the crash.

The NTSB investigation revealed that, despite repeated computer malfunctions, neither CBP nor its contractors had a program in place to ensure that maintenance tasks were performed correctly or that corrective measures were applied. The NTSB investigation also found that while the mishap pilot had only 27 hours of flight time with the Predator B model, and had only been given verbal approval to fly the vehicle accompanied by an instructor pilot, the instructor pilot was not present in the ground control station when the error occurred. NTSB further noted that the verbal approval was not standard practice for CBP.

The incident illustrates that, despite the lack of an onboard pilot, human performance considerations are critical in the design of unmanned aircraft systems. The failure that led to the crash was preventable had the system been designed and operated with certain safeguards in place to compensate for human fallibility. Greater attention to human factors considerations in the design and operation of unmanned aircraft systems and in the development of training programs and operational procedures could improve safety and facilitate the integration of unmanned aircraft in the national airspace system.

Operator Training and Qualification

One particularly important human factors consideration for regulators is a determination regarding training standards and minimum qualification requirements for personnel involved in flight operations. The appropriate training and qualifications may depend in part on the size of the aircraft and the complexity of the system, as a "one size fits all" approach may be inappropriate given the diversity among vehicles, systems, and operational missions.

Currently the U. S. Air Force requires its Predator/Reaper and Global Hawk pilots to be pilot-rated military officers.[25] The other military services do not require drones to be operated by rated pilots, but do require specialized training. It is uncertain how these different approaches may serve as a basis for FAA to develop training requirements and qualification standards for civilian UAV pilots and operators, as military approaches vary considerably and are tailored to specific systems and missions.

One approach would be for FAA to develop basic training and certification standards for operators of small UAVs, while requiring more elaborate training and certification requirements for personnel seeking to operate or maintain large, complex unmanned aircraft systems. The extent to which requirements for UAV personnel may overlap with requirements for piloting or maintaining manned aircraft remains uncertain but could have significant implications for the industry's development. One perceived advantage of unmanned aircraft over manned aircraft is lower operating costs, and mandating that unmanned systems have highly trained pilots and operators with specialized certifications could reduce that potential cost advantage.

Another issue may involve medical certification. Medical conditions or poor eyesight preclude some individuals from obtaining pilot certification, and the rigors of the flight environment demand stringent medical requirements. Given the significant differences in the work environment and the potential consequences of medical problems in an aircraft compared to inside a ground control station, FAA may choose to relax some medical requirements to allow certain individuals that would otherwise not be medically fit to fly to operate unmanned aircraft.

Addressing Potential Security Risks

In addition to safety risks, the operation of civilian unmanned aircraft in domestic airspace raises potential security risks, including the possibility that terrorists could use a drone to carry out an attack against a ground target. It is also possible that drones themselves could be targeted by terrorists or cybercriminals seeking to tap into sensor data transmissions or to cause mayhem by hacking or jamming command and control signals.

Drone Weapons

Terrorists could potentially use drones to carry out small-scale attacks using explosives, or as platforms for chemical, biological, or radiological attacks. In September 2011, the FBI arrested Rezwan Ferdaus, a U. S. citizen from Ashland, MA, charging him in a terrorist plot to attack the Pentagon and the Capitol using large model aircraft packed with high explosives. While the small payload of the model aircraft may have limited the lethality of the explosions, Ferdaus planned to recruit others to use assault rifles to target people fleeing the Pentagon after the drone attack.[26]

In July 2012, Ferdaus pleaded guilty to attempting to provide material support to terrorists and attempting to damage and destroy federal buildings by means of an explosive in a plea agreement under which additional charges were dropped. The incident has raised specific concerns about potential terrorist attacks using unmanned aircraft, although the payload capacities of small UAVs would limit the damage these attacks could inflict using only conventional explosives.

Intentional Hacking and Signal Jamming

Additionally, unmanned aircraft command and control links could potentially be intentionally jammed or hacked resulting in a loss or hostile takeover of control. For example, Todd Humphreys, an assistant professor at the University of Texas at Austin, demonstrated a remote hijacking of an unmanned aircraft by GPS guidance signals. In congressional testimony, he warned that advances in software-defined radio and the availability of GPS signal simulators may provide average hackers with the capability to interfere with unmanned aircraft operations.[27]

Humphreys recommended that non-recreational civilian unmanned aircraft weighing more than 18 pounds be required to have spoof-resistant navigation systems. More broadly, he recommended that GPS-based timing and navigation systems used in national critical infrastructure also be required to be spoof-resistant. He noted that while "[t]here is no quick, easy, and cheap fix for the civil GPS spoofing problem...reasonable, cost-effective spoofing defenses exist which, if implemented, will make successful spoofing much harder."[28] As a long-range solution, he further recommended that the Department of Homeland Security commit to funding the development and implementation of methods for performing cryptographic authentication of GPS signals, or at least for the augmented GPS signals used for civil aviation.

Additionally, GAO noted that "low cost devices that jam GPS signals are prevalent."[29] However, it concluded that the issue could be mitigated by using additional navigation systems that do not rely on GPS, and/or by encrypting communications and telemetry signals.

Availability and Protection of Radiofrequency Spectrum

GAO also noted that the lack of dedicated radiofrequency spectrum for UAS operations raises the risk of lost communications links due to either unintentional or intentional signal interference.[30] Issues regarding dedicated radiofrequency spectrum for unmanned aircraft, which sometimes involve satellite control links, are typically addressed in an international forum to assure global harmonization. The appropriate forum for such determinations is International Telecommunication Union (ITU), the United Nations agency responsible for global information and communications technologies.

In December 2009, ITU issued guidance on spectrum requirements for safe operation of unmanned aircraft in non-segregated or integrated airspace. It concluded that unmanned aircraft operations will require radiofrequency spectrum for air traffic control, vehicle command and control, and sense and avoid capabilities.[31] In November 2011, it published specific guidance on the technical feasibility and advantages and disadvantages of several candidate frequency bands for beyond line of sight communications, command, and control for unmanned aircraft systems.[32] It did not formally consider or propose regulatory methods for how these frequency bands might be allocated, which would be left up to individual regulatory authorities such as the Federal Communications Commission (FCC) in the United States.

FCC has addressed radiofrequency licensing for UAVs on a case-by-case basis, much as FAA has done for certifying drone flight operations. With continued demand for spectrum from mobile broadband providers, securing dedicated radiofrequency spectrum for unmanned aircraft operations is likely to be an ongoing challenge. Moreover, the protection of radiofrequency signals to prevent potential hacking and spoofing through encryption and other methods has not yet been adequately addressed and remains a threat to the security of unmanned aircraft systems operations.

Responsibility for Security Issues

Under the Aviation and Transportation Security Act (ATSA, P.L. 107-71), responsibility for aviation security was transferred from FAA to the newly formed Transportation Security Administration (TSA) in 2001. TSA has not specifically addressed the potential security concerns arising from the operation of drones in domestic airspace.

With regard to matters involving hacking and signal jamming, responsibility for the regulation and oversight of system security is far more complex. While TSA maintains responsibility for airspace security, FAA has regulatory authority over the certification of air vehicles. FAA has not issued certification standards for unmanned aircraft and systems, but future FAA standards may include hardware and software security and reliability assurance requirements.

FAA has worked closely with RTCA and the European Organization for Civil Aviation Equipment (EUROCAE) to define software or information assurance criteria for aircraft systems, and RTCA has released specific guidance on software considerations for airborne systems and equipment. FAA has issued advisory information for aircraft and aircraft equipment manufacturers regarding systems security and software assurance. This information indicates that the RTCA document is only one guideline manufacturers may utilize in applying systems engineering principles to ensure that aircraft component hardware and software are reliable and robust against hacking, signal interference, and other threats.[33]

Additionally, the National Information Assurance Partnership (NIAP), a partnership between the National Security Agency (NSA) and the National Institute of Standards and Technology (NIST), has established common criteria for information technology systems evaluation. Systems meeting the highest levels in the testing process are regarded as having "high robustness," meaning that they have stringent failure protections and rigorous security measures. Some avionics systems for airliners are now being tested using these common criteria methods, although such testing is not currently required. As the issue of unmanned aircraft systems certification is addressed, specific requirements for hardware and software security could be put in place. Additionally or alternatively, drone manufacturers may voluntarily apply various systems engineering and information assurance methods, such as those specified in the RTCA guidance or in the NIAP common criteria methodology, to protect systems against various information security threats.

As with many aspects of unmanned systems, a "one size fits all" approach may not be suitable, and FAA and TSA may choose to adopt a risk-based approach in addressing security matters related to unmanned aircraft operations. Under a risk-based approach, larger, more complex unmanned aircraft may be required to adopt more specific systems security and software and information assurance requirements, while more basic operational security procedures may suffice for small UAVs weighing less than 55 pounds, for example.

In general, a comprehensive approach to security issues pertaining to unmanned aircraft systems operations will likely require close coordination between TSA, FAA, FCC, manufacturers, and operators to address security issues pertaining to systems hardware and software, radiofrequency communications, and flight operations.

Airspace Restrictions

Besides safety and security, the potential impact of drone operations on airspace accessibility for other aircraft has been raised as a specific concern, particularly among operators of smaller general aviation aircraft.

One significant difference between military unmanned aircraft and unmanned aircraft operated by other public agencies or by civilian users is that the military, for many years, has negotiated with FAA to set aside airspace designated for military training, testing, and other purposes. Military airspace consists of military operations areas, military training routes, test ranges, and other restricted or prohibited airspace. Military operations areas and training routes are typically located in remote areas. While not set aside exclusively for military use, these areas and routes are charted and documented, allowing other airspace users, especially smaller general aviation aircraft, to be alert for military flight operations, including operations involving unmanned aircraft. The military has also worked with FAA to set aside specific restricted and prohibited airspace (e.g., above test sites and bombing ranges) to accommodate operations that pose greater risks to other air traffic. This airspace is charted and documented to allow pilots to avoid these areas when restrictions designating them as off limits to non-military users are in effect.

Airspace restrictions can serve as an effective tool for mitigating risks to civilian air traffic. However, their use concerns some airspace users because they pose inconveniences and sometimes raise additional safety issues. With additional restrictions, civilian pilots may be forced to fly more circuitous routes, leaving pilots with fewer options to avoid bad weather or plan for fuel stops. The Aircraft Owners and Pilots Association (AOPA), which represents more than 400,000 general aviation pilots and aviation enthusiasts across the United States, has been particularly critical of the use of restricted airspace designations to separate unmanned aircraft operations from other air traffic. AOPA has asserted that unmanned aircraft operations should not have a negative impact on general aviation operations and should not require special airspace designations, such as restricted airspace, for their operation.[34] It, however, generally supports the integration of unmanned aircraft so long as they do no harm to current manned operations.[35]

AOPA has been particularly critical of FAA's issuance of temporary flight restrictions (TFRs) to separate CBP drone operations for border surveillance from other air traffic. Testifying before the House Subcommittee on Aviation at a March 2006 hearing, AOPA Executive Vice President for Government Affairs Andrew Cebula stated that "AOPA believes that the use of 'temporary' large-scale flight restrictions for yearlong UAV operations is not appropriate and the FAA needs to fully explore the alternatives available to allow CBP (or any other agency for that matter) to secure our borders without impacting the aviation community."[36]

INDUSTRY INITIATIVES

The unmanned aircraft industry, collectively consisting of unmanned aircraft systems and system component manufacturers and system operators, has been cognizant of concerns regarding the operation of unmanned aircraft for civilian uses in domestic airspace. Collectively, the views of drone developers and operators worldwide are represented by the Association for Unmanned Vehicle Systems International (AUVSI). AVUSI has addressed many of the public concerns surrounding UAV operations in a voluntary industry code of conduct issued in July 2012, which is summarized in the text box below.

Voluntary Industry Code of Conduct for Drone Operations

In July 2012, the Association for Unmanned Vehicle Systems International (AUVSI) issued a voluntary industry code of conduct for the operation of unmanned aircraft systems. The code of conduct addresses concerns raised by the public and regulators regarding safety, professionalism, and respect.

With regard to safety, operators adopting the code commit to

- operating in a manner that will not present undue risk to persons or property on the surface or in the air;
- ensuring that vehicles are piloted by competent and properly trained individuals; and
- ensuring that flights are conducted only after thorough risk assessments, including consideration of weather conditions, potential failures and possible consequences, crew fitness for flight operations, airspace issues, applicable aviation regulations, frequency spectrum requirements, and the reliability, performance, and airworthiness of the aircraft.

With regard to professionalism, adopters of the code commit to

- complying with all federal, state, and local laws, ordinances, covenants, and restrictions pertaining to unmanned aircraft operations;
- operating systems as responsible members of the aviation community;
- being responsive to public needs;
- cooperating fully with federal, state, and local authorities in response to emergency deployments, mishap investigations, and media relations; and

- establishing contingency plans for all anticipated non-normal events and sharing these plans openly with all appropriate authorities.

Finally, with regard to respect, operators adopting the code commit to

- respecting the rights of other airspace users;
- respecting the privacy of individuals;
- respecting the concerns of the public regarding unmanned aircraft operations; and
- supporting improved public awareness and education regarding the operation of unmanned aircraft.

Source: Association for Unmanned Vehicle Systems International (AUVSI), Unmanned Aircraft System Operations Industry "Code of Conduct," Arlington, VA.

EYES IN THE SKY: SENSOR PAYLOADS

A central issue surrounding the use of unmanned aircraft is the airborne sensors, particularly imaging sensors, that are mounted to the underbellies of these vehicles to gather data and collect images of the earth below. While these sensors may be utilized for a variety of beneficial applications, they have also raised considerable concerns regarding their potential intrusiveness, despite the aforementioned industry commitments to respect the privacy of individuals (see "Concerns over Privacy and Intrusiveness" below).

Cameras and Electro-Optical Imagers

Typical cameras and sensors placed on unmanned aircraft can collect data across a broad range of the electromagnetic spectrum, both within and beyond the range of human vision. They can provide aerial surveillance capabilities and spectral data under a wide range of lighting and atmospheric conditions. In many cases, small UAVs used in surveillance applications are simply equipped with commercial, off-the-shelf digital still photograph or video cameras. Usually, for aerial work, high-end digital cameras with image stabilization are needed to provide detailed resolution from a distance and avoid motion blur from the movement and vibration of the aircraft and the camera platform. UAVs used for detailed earth imaging are fitted with more advanced high resolution cameras. These cameras may have advanced optical features, including very large digital imaging arrays, large zoom lenses, and advanced image stabilization capabilities, allowing them to take highly detailed aerial photographs.

Captured videos and images may be transmitted for further processing or, in some cases, may be processed onboard by computer vision and video analytic software. Analytics may include tools such as scene motion detection and analysis, object detection capabilities, license plate readers, and, possibly in some cases, facial recognition software.

Drones may also be equipped with electro-optical sensors that are specifically designed for aerial operations. With regard to their imaging capabilities, electro-optical imagers are similar to high-end consumer digital cameras and typically take either still or video images within the visual electromagnetic spectrum, similar to human visual capabilities. Thus these sensors, like cameras, provide eyes in the sky when operated in good daylight and twilight visual conditions, when not obscured by clouds or fog.

Infrared Sensors

Cameras and electro-optical sensors cannot provide quality images during nighttime operations or when operating in low visibility, such as in clouds or in fog. Therefore, to augment electro-optical imaging capabilities and allow for imaging under a broader range of light conditions, sensor platforms often also include two-dimensional infrared (IR) imaging arrays. Since the longer wavelengths in the IR bands are not absorbed or attenuated by clouds or fog and are emitted by warm objects at night, IR sensors are capable of providing images beyond the capabilities or normal human sight. In the far infrared spectrum, passive millimeter wave sensors provide thermal imaging capabilities particularly for night vision and thermal imaging applications. The thermal sensing properties of IR sensors make them well suited for law enforcement and border security applications, for example, for providing eyes in the sky for foot pursuits. IR sensors are also important for command and control of UAVs operated beyond line of sight, giving operators improved situational awareness at night and in poor visibility.

Electro-optical and infrared sensors, collectively known as EO/IR sensors, are often housed together in a turret pod with gyro-stabilized gimbals that allow the sensors to move, either at the direction of a preprogrammed mission plan or from commands issued by a sensor operator remotely controlling the device (see Figure 2).

Source: Customs and Border Protection (CBP).

Figure 2. An Electro-Optical/Infrared (EO/IR) Sensor Housed in a Pod Mounted on an MQ-9 Predator B/Reaper Drone Operated by Customs and Border Protection (CBP).

Synthetic Aperture Radar

In addition to passive EO/IR sensors, aircraft are sometimes fitted with active synthetic aperture radar (SAR) antennas, which measure reflections of emitted microwaves to generate sensor imagery. Real time SAR imagery can penetrate through rain, fog, smoke and dust, providing surveillance capabilities in poor weather and in maritime and desert environments that can augment or replace IR and millimeter wave sensors. Some civilian applications of SAR include maritime search and rescue in adverse conditions, fire line tracking in smoke, iceberg detection and tracking, and oil spill monitoring.[37] While SAR platforms had only been available for large unmanned aircraft until recently, in

2008, defense contractors demonstrated a 2-pound SAR payload for small UAVs that was made commercially available in 2010.[38]

Specialized Sensors

Although cameras, EO/IR, and radars are the most typical types of sensors for UAV applications, unmanned aircraft can be equipped with other sensors for special applications. For example, chemical, biological, or radiological sensors may be used to collect air samples in hazardous environments. NOAA has equipped its UAVs with instruments to collect and analyze air samples.

CONCERNS OVER PRIVACY AND INTRUSIVENESS

While unmanned aircraft may carry a wide variety of sensor payloads for diverse applications, cameras and imaging sensors have raised particular concerns among privacy advocates who fear that widespread civilian unmanned aircraft operations could lead to abuses in monitoring, tracking, and surveilling people throughout the courses of their daily lives. Civil liberties and privacy groups including the American Civil Liberties Union (ACLU) and the Electronic Privacy Information Center (EPIC) argue that aerial drones pose a unique threat to privacy.[39] Proponents of drones have countered these arguments by underscoring the many potential benefits of domestic drone use. For example, aerial imaging of farm crops across multiple spectral bands can provide data on crop conditions and may provide critical information on the impacts of drought conditions or other factors affecting crop health. Drone proponents also contend that while unmanned aircraft have been singled out in public policy debate, the sensor technologies, imaging systems, and data collection capabilities in question are not unique to unmanned aircraft, as these devices could just as easily be installed on manned aircraft, or on utility poles, cell phone towers, and tall buildings.[40]

While the main focus of current public policy debate regarding drone use centers on the law enforcement and government use of UAVs for surveillance purposes, additional concerns have been raised regarding the commercial use of drones, including their potential use by marketing firms, private investigators, and paparazzi. While such uses may be considered intrusive by many, Fourth Amendment concerns would not specifically apply to data collections that are not conducted by or on behalf of governmental entities.

Besides tracking and surveillance activities that raise specific privacy concerns, civilian unmanned aircraft operations may raise additional concerns regarding their potential intrusiveness. Various potential commercial uses, such as aerial advertising (using drones to tow banners or display other targeted visual advertisements), may not raise privacy concerns per se, but may nonetheless be regarded by some as intrusive and a nuisance because of aircraft noise and visual distraction.

FAA's authority over specific uses for civilian unmanned aircraft appears limited. As a matter of law, aircraft have privileged access to navigable airspace. U. S. law generally maintains that equal access and availability of navigable airspace to all users are basic principles underlying the national airspace system, so long as safety and national security are not compromised. Notably, 49 U. S.C. Section 40101(c) (2) specifies that FAA, in carrying out its authority to regulate aviation, shall consider the public right of freedom of transit through the navigable airspace.

Navigable airspace is defined in statute as that airspace above minimum flight altitudes prescribed by regulation, including airspace needed to ensure safety during takeoff and landing.[41] Away from airports, minimum altitudes defining navigable airspace are largely predicated on safety concerns and depend on characteristics of the aircraft and risks posed to persons and property on the ground. In general, navigable airspace is considered to be 500 feet above ground level and higher, except in congested areas where the minimum safe altitude rises to 1,000 feet above the highest obstacle within a 2,000-foot radius of the aircraft's position at any given time.[42] In sparsely populated areas and over water, navigable airspace can be considered to be much lower, so long as operators assure that they can make an emergency landing after an engine failure without undue hazard to persons or property on the ground. Moreover, helicopters, powered parachutes, and powered hang gliders can be operated below established minimums so long as operations are conducted without hazard to persons or property on the ground, and so long as helicopters follow certain prescribed routes and altitudes when defined, usually for noise abatement purposes. Minimum altitudes for drones are yet to be defined. However, unregulated model aircraft are generally permitted below 400 feet, outside of navigable airspace for most manned aircraft.

Small low-flying drones may pose particular concerns regarding intrusiveness, especially if they become widely proliferated in urban and suburban areas. For example, it is imaginable that drones displaying advertising material might hover noisily above a beach, or that a magazine might deploy a drone equipped with high-resolution cameras to photograph a celebrity's garden party. As FAA is primarily concerned with balancing safety concerns with its obligation to facilitate access to navigable airspace, it is likely to tread lightly with regard to matters pertaining to the regulation of specific drone uses based upon their perceived intrusiveness.

Although FAA may be reluctant to address such issues, matters pertaining to the use of domestic airspace generally fall exclusively within its purview. 49 U. S.C. §40103(a)(1) specifies that the federal government has exclusive sovereignty of domestic airspace, thus significantly limiting state and local governments from promulgating laws, regulations, or ordinances affecting access to the national airspace system. This may make it particularly difficult for state and local governments to restrict or regulate the use of drones, including imposing limitations on noise.

On the federal level, landowner rights with regard to the navigable airspace above private property are similarly restricted. The Supreme Court has stated that "the air is a public highway, as Congress has declared. Were that not true, every transcontinental flight would subject the operator to countless trespass suits. Common sense revolts at the idea."[43]

While the concept of the air as a public highway has largely overturned the ancient Roman concept that the owner of a land also holds title to the air above, it does not preclude civilians from potentially seeking civil remedies in the courts against aircraft operators flying overhead or in close proximity to an individual's land. This may potentially include civil action against private or commercial drone operators, if damage or interference can be demonstrated. It may also include claims against federal drone operators if the operations are considered to constitute a taking of property by the government because the aviation activity results in a loss of use or a loss of enjoyment of the property. The Fifth Amendment, among other things, protects citizens from a taking of private property for public use without just compensation.

In these matters, case law relating to aircraft noise may have particular relevance to potential landowner actions in response to nuisances and interference caused by drone operations. Nuisance suits pertaining to aviation noise have allowed recovery for annoyance, inconvenience, discomfort, and mental and emotional distress, as well as for property damage. Additionally, citizens have been successful in actions against the federal government under the "just compensation" clause of the Fifth Amendment in cases where federal aviation operations, usually from military airbases or military training areas, have interfered with the use and enjoyment of private land, thereby constituting a "taking" of the land without any formal exercise of powers of eminent domain. To the extent that case law regarding aircraft noise from the early days of civil jet air transportation may bolster potential legal challenges to drone operations, history suggests that the introduction of civilian drones in the national airspace system is likely to be a highly contentious matter.

End Notes

[1] Unmanned aircraft systems (UASs) refer to aircraft that fly without onboard pilots along with ground control stations, networks, and personnel needed to operate these aircraft. Unmanned aerial vehicles (UAVs) refer to the unmanned aircraft in these systems, which are often referred to as drones in the media. The terms UAVs, unmanned aircraft, and drones are used interchangeably throughout this report.

[2] Kettering Aerial Torpedo Bug, *Fact Sheet*, National Museum of the Air Force: Dayton, Ohio, available at http://www.nationalmuseum.af.mil/factsheets/factsheet.asp?id=320.

[3] For further information on military and intelligence UASs see CRS Report R42136, *U. S. Unmanned Aerial Systems*, by Jeremiah Gertler.

[4] For additional information see the Academy of Model Aeronautics (AMA) website at http://www.modelaircraft.org.

[5] Federal Aviation Administration, *Model Aircraft Operating Standards*, Advisory Circular AC 91-57, June 9, 1981, Washington, DC.

[6] Teal Group Corporation, *Teal Group Predicts Worldwide UAV Market Will Total $89 Billion in Its 2012 UAV Market Profile and Forecast*, available at http://tealgroup.com/index.php/about-teal/teal-group-in-the-media/3/79-teal-grouppredicts-worldwide-uav-market-will-total-89-billion-in-its-2012-uav-market-profile-and-forecast.

[7] VisionGain, *The UAV Payload and Subsystem Market 2012-2022*, July 19, 2012, available at http://www.visiongain.com/Report/857/The-UAV-Payload-and-Subsystems-Market-2012-2022; VisionGain, *The UAV Flight Training & Simulation Market 2012-2022*, May 17, 2012, available at http://www.visiongain.com/Report/823/ The-UAV-Flight-Training-Simulation-Market-2012-2022.

[8] U. S. Government Accountability Office, *Unmanned Aircraft Systems: Use in the National Airspace System and the Role of the Department of Homeland Security*, Statement of Gerald L. Dillingham, Ph.D., Director, Physical Infrastructure Issues, Before the Subcommittee on Oversight, Investigations, and Management, Committee on Homeland Security, House of Representatives, July 19, 2012, GAO-12-889T.

[9] Federal Aviation Administration, *FAA Aerospace Forecasts FY2012-2032*, p. 57.

[10] 49 U. S.C. §40125 specifies the qualifications for public aircraft status.

[11] P.L. 107-76, Section 709c(6), italics added.

[12] Federal Aviation Administration, "Unmanned Aircraft Operations in the National Airspace System," *Federal Register*, 72(39), February 13, 2007, 6689-6690.

[13] Federal Aviation Administration, *Fact Sheet: Unmanned Aircraft Systems (UAS)*, Updated July 2011, available at http://www.faa.gov/about/initiatives/uas/media

[14] David Hughes, "Civil Use of Unmanned Aircraft Systems Could Grow Rapidly," *Aviation Week and Space Technology*, February 12, 2007, p. 49.

[15] Katherine McIntire Peters and Beth Dickey, "Droning On," Government Executive, October 15, 2004, pp. 68-76.

[16] Federal Aviation Administration, *Fact Sheet: Unmanned Aircraft Systems (UAS)*, Updated July 2011, available at http://www.faa.gov/about/initiatives/uas/media

[17] *Drone pilots* refers to ground personnel responsible for direct flight control of the unmanned vehicle whereas the term *systems operators* refers to personnel responsible for operating sensor payloads or other aircraft systems, such as fuel systems, that do not involve direct manipulation of flight controls or flight guidance systems. Other safety critical personnel may include maintenance personnel and equipment technicians that service the air vehicle, the ground station or command and control linkages between the vehicle and the ground station.

[18] Based on 2010 total combined aircraft operations at towered airports reported in Federal Aviation Administration, *FAA Aerospace Forecasts FY2012-2032*. This does not include activity at non-towered airports and, therefore, understates the total number of daily flights, for which there is no definitive count.

[19] U. S. Government Accountability Office, *Unmanned Aircraft Systems: Use in the National Airspace System and the Role of the Department of Homeland Security*, Statement of Gerald L. Dillingham, Ph.D., Director, Physical Infrastructure Issues, Before the Subcommittee on Oversight, Investigations, and Management, Committee on Homeland Security, House of Representatives, July 19, 2012, GAO-12-889T.

[20] "Sense and avoid" refers to technologies and capabilities allowing unmanned aircraft to reliably detect other air traffic and maneuver away from such traffic in a manner that adequately mitigates collision risks.

[21] RTCA was organized in 1935 as the Radio Technical Commission for Aeronautics. Today it operates as a private, not-for-profit corporation. Its recommendations are used by FAA as a basis for policy, program, and regulatory decisions.

[22] Roland E. Weibel and R. John Hansman, *Safety Considerations for Operation of Unmanned Aerial Vehicles in the National Airspace System*, MIT International Center for Air Transportation, Department of Aeronautics and Astronautics, Massachusetts Institute of Technology, Cambridge, MA, March 2005, Report No. ICAT-2005-1.

[23] Ibid.

[24] National Transportation Safety Board, NTSB Identification: CHI06MA121, Aviation Accident Database & Synopses, Probable Cause Approval Date: October 31, 2007. Available at http://www.ntsb.gov/aviationquery/ brief.aspx?ev_id=20060509X00531&key=1.

[25] See CRS Report R42136, *U. S. Unmanned Aerial Systems*, by Jeremiah Gertler.

[26] Milton J. Valencia, "Ashland Man Faces Terrorism Charges," *Boston Globe*, September 29, 2011.

[27] Todd Humphreys, *Statement on the Vulnerability of Civil Unmanned Aerial Vehicles and Other Systems to Civil GPS Spoofing*, Submitted to the Subcommittee on Oversight, Investigations, and Management of the House Committee on Homeland Security, July 19, 2012.

[28] Ibid. In this context, spoofing refers to transmitting false navigation signals that could cause a drone or aircraft to veer off course.

[29] U. S. Government Accountability Office, *Unmanned Aircraft Systems: Use in the National Airspace System and the Role of the Department of Homeland Security*, Statement of Gerald L. Dillingham, Ph.D., Director, Physical Infrastructure Issues, Before the Subcommittee on Oversight, Investigations, and Management, Committee on Homeland Security, House of Representatives, July 19, 2012, GAO-12-889T.

[30] Ibid.

[31] International Telecommunication Union, ITU-R Radiocommunication Sector of ITU, Characteristics of Unmanned Aircraft Systems and Spectrum Requirements to Support Their Safe Operation in Non-Segregated Airspace, Report ITU-R M. 2171, December, 2009, Geneva, Switzerland.

[32] International Telecommunication Union, ITU-R Radiocommunication Sector of ITU, Frequency Sharing Between Unmanned Aircraft Systems for Beyond Line of Sight Control and Non-Payload Communications Links and Other Existing and Planned Services in the Frequency Bands 13.25-13.40 GHz, 15.4-15.7 GHz, 22.5-22.55 GHz and 23.55- 23.60 GHz, Report ITU-R M. 2230, November, 2011, Geneva, Switzerland.

[33] Federal Aviation Administration, Advisory Circular 20-115B—Radio Technical Commission for Aeronautics, Inc. Document RTCA/DO-178B, January 11, 1993.

[34] Aircraft Owners and Pilots Association, *Regulatory Brief Unmanned Aircraft Systems*, Updated January 31, 2008, Frederick, MD, available at http://www.aopa.org/whatsnew/regulatory/unmanned.html.

[35] Benet J. Wilson, "Integrating Manned, Unmanned Vehicles in Airspace System," Aircraft Owners and Pilots Association, Frederick, MD, available at http://www.aopa.org/advocacy/articles/2012/120608intregating-mannedunmanned-vehicles-in-airspace-syste.html.

[36] Aircraft Owners and Pilots Association, Statement of Andrew V. Cebula, Executive Vice President, Government Affairs Concerning Unmanned Aerial Vehicles in the National Airspace System, Before the Committee on Transportation and Infrastructure, Aviation Subcommittee, U. S. House of Representatives, March 29, 2006.

[37] Mike Hanlon, "ScanEagle UAV Gets Synthetic Aperture Radar (SAR), *GizMag*, March 18, 2008, available at http://www.gizmag.com/scaneagle-uav-gets-synthetic-aperture-radar

[38] Ibid., "Insitu Announces Availability of NanoSAR—Synthetic Aperture Radar Payload." *BusinessWire*, February 23, 2010, available at http://www.businesswire.com/news/home/20100223005511/en.

[39] Jay Stanley and Catherine Crump, *Protecting Privacy From Aerial Surveillance: Recommendations for Government Use of Drones,"* American Civil Liberties Union, New York, NY, December 2011; Testimony and Statement for the Record of Amie Stepanovich, Association Litigation Counsel, Electronic Privacy Information Center, Hearing on "Using Unmanned Aerial Systems Within the Homeland: Security Game Changer?" Before the Subcommittee on Oversight, Investigations, and Management of the U. S. House of Representatives Committee on Homeland Security, July 19, 2012.

[40] For further information on the legal implications of domestic drone use, including Fourth Amendment and other privacy concerns, see CRS Report R42701, *Drones in Domestic Surveillance Operations: Fourth Amendment Implications and Legislative Responses*, by Richard M. Thompson II.

[41] 49 U. S.C. §40102(a)(32).

[42] See 14 C.F.R. §91.119 Minimum safe altitudes: General.

[43] United States v. *Causby*, 328 U. S. 256, 260-61 (1946).

In: Domestic Drones
Editor: Donovan Ambrosio

ISBN: 978-1-63321-456-9

Chapter 4

REPORT TO CONGRESS ON FUTURE UNMANNED AIRCRAFT SYSTEMS TRAINING, OPERATIONS, AND SUSTAINABILITY*

U. S. Department of Defense

Preparation of this report/study cost the
Department of Defense a total of approximately
$17,000 in Fiscal Years 2011 — 2012. RefiD: 7-3C47E5F

REPORTING REQUIREMENT

This report is being provided to the congressional defense committees as requested in House Armed Services Committee Report 111-491, accompanying the Ike Skelton National Defense Authorization Act for Fiscal Year 2011.

TASK: Report on Future Unmanned Aerial Systems Training, Operations, and Sustainability. The rapid growth of UAS inventories to meet operational demands raises a number of questions concerning the military services' ability to support these inventories in the near-and long-term. In particular, to support their UAS inventories, the military services must train sufficient numbers of personnel to operate and maintain the aircraft, provide adequate

* This is an edited, reformatted and augmented version of a report issued April 2012.

facilities and other infrastructure to sustain them, and provide sufficient access to airspace and training ranges to train military personnel within the United States and at military bases overseas.

The committee directs the Secretary of Defense to provide a report to the congressional defense committees with its fiscal year 2012 budget request that describes the military services' plans to support their current and planned UAS inventories. The report should, at a minimum, discuss:

1. Current UAS inventory levels and planned UAS inventory levels for each fiscal year through 2017;
2. Plans to supply the number of personnel needed to operate the aircraft and sensor payloads and to perform UAS maintenance;
3. Current and planned UAS basing and other operating locations;
4. Progress made in providing the number of facilities needed for UAS inventories to support operations and training and the funding required for any additional facilities; and
5. The availability of airspace, ranges, and other infrastructure at each planned UAS location, and a description of the steps that the services plan to take to overcome any limitations that adversely impact UAS training.

EXECUTIVE SUMMARY

The Department of Defense (DoD) continues to increase its investment in unmanned aircraft systems (UAS) to meet battlefield commanders' demand for their unique capabilities. The emphasis on long-endurance, unmanned intelligence, surveillance and reconnaissance (ISR) assets — many with strike capabilities — is a direct reflection of recent operational experience and further Combatant Commander demands. This increase in demand has resulted in a large number of UAS capable of a wide range of missions. This large number of fielded UAS has also driven a strong demand for access within the National Airspace System (NAS). This need for airspace access to test new systems, train operators, and conduct continental United States (CONUS)-based missions has quickly exceeded the current airspace available for military operations. The situation will only be exacerbated as units return from overseas contingencies. Currently, DoD UAS operations conducted outside of Restricted, Warning, and Prohibited areas are authorized under a temporary Certificate of Waiver or Authorization (COA) or waiver from the Federal

Aviation Administration (FAA) or under limited conditions outlined in the 2007 DoD-FAA Memorandum of Agreement (MoA). DoD is actively engaged in coordinating efforts on behalf of the Military Departments and Combatant Commands to shorten and simplify the FAA COA process to allow greater unmanned access to the NAS, with direct engagement through the interagency UAS Executive Committee (ExCom). The UAS ExCom is a joint committee composed of senior executives from four member organizations: DoD, FAA, the Department of Homeland Security (DHS), and the National Aeronautics and Space Administration (NASA). The mission of the UAS ExCom is to enable increased and ultimately routine access of Federal UAS engaged in public aircraft operations into the NAS to support operational, training, development, and research requirements of FAA, DoD, DHS, and NASA. DoD is also pursuing ground-based and airborne sense-and-avoid efforts to eventually supplant or significantly reduce the need for COAs. In the future, DoD will continue to utilize Restricted, Warning, and Prohibited areas but will also continue to develop the necessary technologies to access other airspace safely and in accordance with applicable federal aviation regulations.

This document outlines planned force capability growth and forecasted attrition of UAS aircraft through FY 2017; Military Department personnel required for training and operations; personnel and aircraft basing intentions; and required military construction (MILCON) and airspace requirements for bases hosting UAS. Within the report, the Military Departments provide current and planned inventories, personnel requirements to operate and maintain the systems, planned bases and operating locations, and progress with facilities to support inventories. Also, the report addresses the airspace integration challenge through implementation of the DoD Airspace Integration Plan, multi-agency collaboration, and ongoing negotiations with FAA. The Military Departments have a cohesive plan to address basing, funding, and manning in support of forecasted training and operations.

INTRODUCTION

Effective employment of UAS worldwide is an integral part of DoD military operations. UAS operations in the NAS are required to ensure direct mission support to Combatant Commanders to both train and maintain ready forces and pursue operational test activities for UAS. Additionally, UAS are utilized to conduct Homeland Defense/Homeland Security and, when approved by the Secretary of Defense, Defense Support of Civil Authorities

(DSCA) missions (e.g., disaster relief, search and rescue). Accomplishing these missions requires airspace to efficiently train, develop, and support UAS operations. This report describes the Military Departments' UAS inventories, personnel, sustainment, and site plans to support and execute UAS missions from now through FY 2017.

SECTION 1. CURRENT UAS INVENTORY LEVELS AND PLANNED UAS INVENTORY LEVELS FOR EACH FISCAL YEAR THROUGH FY 2017

The following table describes the current UAS program of record inventory levels planned through FY 2017, net of attrition.

Table I. UAS Inventory Levels (FVI2 budgeted inventory with noted exception)

System Designation/Name I		Current	FY 12	FY 13	FY 14	FY 15	FY 16	FY 17
		Air Force						
MQ-1B	Predator	163	152	141	130	121	115	110
MQ-9A	Reaper	70	96	135	167	199	229	256
RQ-4B *	Global Hawk	23	23	15	15	15	15	15
		Army						
RQ-11B	Raven	5394	6294	6528	6717	6921	7074	7074
RQ-7B	Shadow	408	408	408	408	408	408	408
MQ-5B	Hunter	45	45	45	45	45	45	45
MQ-1C	Gray Eagle	19	45	74	110	138	152	152
		Navy						
RQ-4A	Global Hawk	5	5	0	0	0	0	0
MQ-4C	BAMS	0	0	2	2	5	9	13
MQ-8B	FirescoutNTUAV	5	9	14	18	25	32	37
RQ-21A	STUAS	0	1	2	3	4	4	4
	Scan Eagle	122	122	122	122	122	122	122
X-47B	UCAS-D	2	2	2	2	0	0	0
	UCLASS	0	0	0	0	2	2	4
		Marine Corps						
RQ-7 B	Shadow	52	52	52	52	52	52	52
RQ-21A	STUAS	8	8	8	23	48	73	100

* Reflects RO-4B Block 20/40 inventory remaining after FY 2012 (Block 30 cancelled in President's 2013 Budget submission).

SECTION 2. PLANS TO SUPPLY THE NUMBER OF PERSONNEL NEEDED TO OPERATE THE AIRCRAFT AND SENSOR PAYLOADS AND TO PERFORM UAS MAINTENANCE

This section provides manpower planning by the Military Departments for the necessary personnel to operate the aircraft and mission sensor. UAS pilots/operators require initial, continuation, upgrade, and proficiency/currency training sorties in the NAS. Similar requirements apply to sensor operators in their respective mission areas.

Air Force

As of December 16, 2011, the manpower requirements for Remote Piloted Aircraft (RPA) pilots and Sensor Operators (SO) to support 57 MQ-1/9 and 4 RQ-4 Combat Air Patrols (CAPs)[1], including operational, test, and training requirements, as well as appropriate overhead and staff requirements, were:

Table 2. RPA Crew Manpower Requirements

	MQ-1	MQ-9	RQ-4	Total
Pilots	1012	529	155	1696
SO	730	401	63	1194

As of December 16, 2011, the number of trained RPA pilots and SOs available and the resulting personnel shortfall to provide 57 MQ-1/9 and 4 RQ-4 CAPs, including operational, test, and training requirements, as well as appropriate overhead and staff requirements, was:

Table 3. Current RPA Crew Manning Availability

	MQ-1	MQ-9	RQ-4	Total	Current Shortfall
Pilots	726	455	177	1358	-338
SO	610	291	48	949	-245

The temporary shortfalls in aircrew manning were overcome by using a minimum of seven aircrews vice the programmed ten aircrews per CAP and by prioritizing operational, test, and training requirements above overhead and staff requirements.

Beginning on March 30, 2011, the Air Force was tasked to provide additional CAPs to support new contingency operations in Libya and a summer surge in Afghanistan. During the fall and winter of 2011, the Air Force provided 60 MQ-1/9 CAPs and 4 RQ-4 CAPs. The Air Force took the following actions in order to support this additional temporary surge:

a) Stood down a portion of the formal training to provide three CAPs;
b) Mobilized Air Force Reserve (AFR) and Air National Guard (ANG) units to provide two CAPs for 7 months;
c) Resourced ANG CAP with volunteers; and
d) Accelerated the early stand up of two planned CAPs.

These actions impacted the production of trained aircrews, requiring the Air Force to reconstitute the force during FY 2012. On December 16, 2011, the Air Force reduced sourcing to 57 MQ-1/9 CAPs, allowing experienced aircrews to be reassigned to the formal training units at Holloman AFB as instructors. Additionally, the Air Force will increase hiring of contract instructors to augment uniformed instructors. These actions will enable the Air Force to increase RPA aircrew manning to full strength at a sustainable rate.

The FY 2015 manpower requirements for RPA pilots and SOs to provide 65 MQ-1/9 and 8 RQ-4 CAPs, including operational, test, and training requirements, as well as all overhead and staff requirements, are:

Table 4. FY 2015 RPA Crew Requirements

	MQ-1	MQ-9	RQ-4	Total
Pilots	902	858	300	2060
SO	657	647	150	1454

In order to meet this RPA aircrew manpower requirement, the Air Force has implemented two key initiatives. The first initiative created Undergraduate RPA Training (URT) for RPA pilots with the 18X Air Force Specialty Code (AFSC) as well as a distinct training pipeline for RPA SOs with the **I** U AFSC. These programs solve the problem of insufficient capacity in existing pipelines (Undergraduate Pilot Training (UPT) and IN AFSC training) to meet RPA aircrew operational requirements. URT production is planned at 60 for FY 2011, 146 for FY 2012, and 168 in FY 2013-2015, while the 1U training pipeline is planned for 353 in FY 2011, 327 in FY 2012, 255 in FY 2013, and 202 in FY 2014-2016. The 18X and 1U career fields will comprise the majority of the RPA aircrew force structure in the future. Until that time, the Air Force will continue to use traditional pilots with the 11X AFSC and SOs with the IN and IA AFSCs to augment the RPA aircrew requirement.

The second initiative increased the capacity of the MQ-1/9 Formal Training Units (FTUs) in order to meet operational RPA requirements. There is currently one active duty MQ-1 FTU and one active duty MQ-9 FTU as well as a launch and recovery training squadron. Additionally, there is an Air National Guard MQ-1 FTU and an Air National Guard MQ-9 FTU producing pilots in FY 2012. As the Air Force evolves toward an MQ-9 fleet, aircrew production focus will shift from MQ-1 to MQ-9, which will require standing up an additional active duty MQ-9 FTU. The expected capacity of the MQ-1/9 FTUs will be 310 initial qualification MQ-1/9 aircrew and 30 MQ-1 to MQ-9 aircrew conversions in FY 2012 and 360 initial qualification MQ-1/9 aircrew and 40 MQ-1 to MQ-9 aircrew conversions in FY 2013. These training slots support Active/Reserve Component and foreign training requirements. For the RQ-4, there is one active duty FTU that has a capacity of 72 pilots and 36 SOs per year. This RPA pilot training infrastructure and the associated capacity will enable the Air Force to meet the operational RPA aircrew requirement and continue to sustain the enterprise in the future.

Air Force RPA organizational level maintenance utilizes a combination of military and Contracted Logistics Support (CLS) personnel in support of operations. Organizational-level maintenance contractors primarily reside within the MQ-1 community, performing both home station and deployed maintenance actions alongside military maintenance technicians. Military maintenance personnel currently perform the majority of organizational level maintenance within the MQ-9 community. MQ-1/9s require up to 30 military maintenance personnel to stand-up an initial CAP. This number increases up to 65 military maintenance personnel supporting up to five CAPs. MQ-1 contractors gain efficiencies by conducting Cross Utilization Training (CUT) with their personnel enabling the contractor to conduct home station and deployed maintenance operations with a significantly smaller footprint. The RQ-4 also utilizes a combination of military maintenance and CLS personnel to perform the majority of organizational-level maintenance actions. RQ-4s require up to 60 military maintenance personnel per detachment to support operations. As with the MQ-1/9s, contractors gain efficiencies by conducting CUT training with their personnel to reduce footprint.

Original equipment manufacturers currently conduct all Air Force RPA depot-level maintenance actions. A Business Case Analysis (BCA) to determirie a course of action for switching to a more organic depot structure is underway within the MQ-1/9 community. Preliminary findings will be concluded by June 2012. The Life Cycle Sustainment plan for the RQ-4 is complete and awaiting final signatures. A BCA to determine a course of action for switching to a more organic depot structure will follow in the near future.

Air Force Maintenance Career Field Managers conducted Utilization and Training Workshops in April and May of 2011 to determine training requirements for all aircraft maintenance AFSCs with the exception of Weapons. An interim mechanical RPA course (Crew Chief, Engines and Hydraulics technicians) began in August 2011. For the long term, robust mechanical and technical courses are currently under development with implementation of a mechanical course scheduled for August 2012.

Army

The Army uses three Military Occupation Specialties (MOS) to support UAS. Two of these MOS, 15W Operator and 15E Repairer, are for enlisted Soldiers and one, 150U Technician, is for a Warrant Officer. The exception to this is the small RQ-11B Raven systems which are operated by any Soldier qualified through a 10-day flight training course.

The 15W Operator is qualified to fly the aircraft, operate the sensors, and emplace/displace the system. Individual aircraft qualifications are tracked by an additional skill identifier. The 15W is the feeder MOS for the 150U Technician MOS.

The 15E Repairer is responsible for the maintenance of all parts of the UAS, to include the aircraft, ground control stations, data links, and supporting equipment. The 15E, like the 15W, has an additional skill identifier to denote specific system qualifications.

The 150U Warrant Officer provides leadership and expertise in the UAS unit. These individuals interface with their higher headquarters and provide oversight of aviation safety, standardization, and maintenance programs.

Manning numbers are based on Unit Modified Table of Organizational Equipment (MTOE) requirements. The exception to this is the RQ-11B Raven where the number of trained personnel is at the discretion of the owning

Commander. The numbers included in this document for Raven are the minimum required.

Table 5. Current Manpower Requirements

	MQ-IC	MQ-5B	RQ-78	RQ-11B	Total
15E Operator	92	57	1307	N/A	1456
15W Mechanic	52	33	733	N/A	818
150U Warrant	20	12	168	N/A	200
Total	164	94	2208	3596	

Table 6. FY 2015 Manpower Requirements

	MQ-1C	MQ-5B	RQ-7B	RQ-11B	Total
15E Operator	552	57	1448	N/A	2057
15W Mechanic	312	33	828	N/A	1173
150U Warrant	120	12	184	N/A	316
Total	984	94	2460	4614	

Due to the rapid growth of the Army UAS fleet, all three of these MOS have been stressed to maintain pace with demand. To ensure wartime requirements are met, the Army has prioritized the distribution of UAS personnel to units preparing to deploy, followed by new unit fielding and then other units. This prioritization has allowed the Army to meet wartime requirements with well-trained and integrated units.

Navy

The Navy is currently conducting strategic planning for the long term manpower required to operate and maintain its UAS.

Vertical Take off and Landing Tactical Unmanned Aerial Vehicle (VTUAV)-Littoral Combat Ship (LCS)

Detachments that operate from the LCS are known as composite Aviation Detachments (AvDets). These composite AvDets operate and maintain both the MH-60R or MH-60S and one or two MQ-8B Fire Scout aircraft. This minimally manned (4 officers, 19 enlisted) detachment structure cannot be split to operate the different aviation platforms independently. Therefore, the UAS portion of LCS composite AvDet manning is not separate from traditional helicopter manning.

The MH-60/VTUAV composite AvDets will support all LCS seaframes with a 3:3:1 rotation. Under the 3:3:1 plan, three aviation detachments of personnel will use three AvDets to support one LCS deployment. One AvDet will be deployed, one AvDet will be completing training requirements as it prepares to deploy, and one AvDet will be conducting turnover and upkeep training to maintain aviation proficiency having recently returned from a deployment.

At the completion of the fleet response training plan (FRTP) cycle, personnel will return to their squadrons for further assignment, in accordance with the squadron's detachment loading, while MQ-8B airframes will be returned to the Contract Logistics Site (CLS) base. Personnel will be managed within squadrons to ensure operational exposure and experience is gained in both manned and unmanned assignments to enable a quality spread of personnel, and to attain personal career progression milestones. This will promote VTUAV community integration and ensure the community has a flexible manpower base to draw upon when supporting various detachment configurations.

VTUAV-Special Operations Forces (SOF) ISR

Navy is evaluating the manning structure needed to provide a VTUAV-only aviation detachment. The leading proposal is to develop unmanned detachments, or "UDets" from HQU-10, the VTUAV fleet replacement squadron. Nine detachments would be needed to support three constantly deployed UDets, totaling approximately 336 additional personnel. Alternately, a separate expeditionary VTUAV squadron could be organized to perform the same function but at higher personnel cost due to the lack of synergy with an existing unit.

BAMS

Personnel that will operate and maintain the MQ-4C BAMS will transition from the existing Maritime Patrol (P-3C Orion) community. The existing community will transition from the P-3C to the P-8A Poseidon and the MQ-4C. At full operational capability in FY 2020, it is estimated that 866 personnel will be needed for five worldwide BAMS orbits.

Marine Corps

U. S. Marine Corps (USMC) Shadow and STUAS

Three active-duty and one reserve-duty Marine Unmanned Aerial Vehicle (VMU) squadrons are manned at or near the required Table of Organization (T/O) manning level of 193 Marines and Sailors. Each VMU is comprised of a headquarters element (40 personnel) and three individual RQ-7B detachments, each comprised of 51 Marines. Each RQ-7B detachment is capable of independent operations, and contains all necessary intelligence, communications, flight operations, and maintenance personnel. In 2012, an additional 81 Marines to support nine RQ-21A STUAS systems will begin arriving at each active-duty VMU squadron. Each RQ-21A STUAS detachment will contain nine Marines and be capable of independent UAS flight operations with required host unit support.

SECTION 3. CURRENT AND PLANNED UAS BASING AND OTHER OPERATING LOCATIONS

The rapid increase in fielded UAS has created a strong demand for access within the NAS and international airspace. The demand for airspace to test new systems and train UAS operators has quickly exceeded the current airspace available for these activities. Figure 1 below shows the projected number of DoD UAS locations in the next 6 years, many without access to airspace compatible for military operations under the current regulatory environment.

NAS access for UAS is currently limited primarily due to regulatory compliance issues and interim policies. DoD UAS operations conducted outside of restricted, warning, and prohibited areas are authorized only under a (temporary) COA from the FAA. The COA process is adequate for enabling a small number of flights but does not provide the level of airspace access necessary to accomplish the wide range of DoD UAS missions at current and projected operational tempos (OPTEMPOs). This constraint will only be exacerbated as combat operations shift from abroad and systems return to U. S. locations.

If DoD UAS do not have direct access to Restricted and Warning Areas (e.g., airfield located within a restricted area), a COA is required. Obtaining a COA requires a significant amount of time and resources — both to complete an application and to work through the FAA approval process. The Military Departments currently have 88 active COAs at various locations around the country, most of which provide access to a restricted or warning area. Many restricted areas are small in size and will only accommodate a smaller sized UAS.

Figure 1. Representative DoD UAS Locations by 2017.

Table 7 lists the Departments' 110 potential UAS basing locations and the UAS likely to fly at that location.

Table 7. Planned DoD VAS Locations by State/Territory

Location	Service	UAS
ALABAMA		
f mt McClellan	ARNG	RQ11B
	ARNG	RQ7B
Redstone	USA	RQ11B
	USA	RQ7B
ALASKA		
Fort Greely	USA	RQ11B
	USA	RQ7B

Location	**Service**	**UAS**
Fort Richardson	USA	RQ11e
	USA	RQ7B
Fort Wainwright	USA	RQ11B
	USA	RQ7B
ARIZONA		
Cochise College	USA	MQ5
Florence	ARNG	RQ11B
Fort Huachuca	USA	MQ1B
	USA	MQ1C
	USA	MQ5
	USA	RQ7B
	USMC	RQ7B
	USMC	RQ21A
MCAS Yuma	USMC	RQ7B
	USMC	RQ21A
Whetstone	ARNG	RQ7B
Yuma Proving Ground	USN	MQ8
ARKANSAS		
Fort Chaffee	ARNG	RQ11B
	ARNG	RQ7B
Little Rock	ARNG	RQ1113
CALIFORNIA		
Beale AFB	USAF	RQ4
	USN	RQ4A GAMS
Camp Morena	USSOCOM	PUMA AE
	USSOCOM	RQ11B
	USSOCOM	Wasp
NAWS China Lake	USN	MQ8
Camp Roberts	ARNG	RQ11B
	ARNG	RQ7B
	USSOCOM	Scan Eagle
El Mirage	USA	MQ1B
	USA	MQ1C
Filmore	USA	Wasp
Filmore	USA	PUMA AE

Table 7. (Continued)

Location	Service	UAS
	USSOCOM	Wasp
	USSOCOM	Puma AE
Fort Irwin	USA	RQ11B
	USA	RQ7B
Gray Butte	USAF	MQ1B
	USAF	MQ9A
MCAGCC 29 Palms	USMC	RQ-11B
	USMC	RQ-7B
	USMC	Scan Eagle
	USMC	RQ21A
MCB Camp Pendleton	USMC	RQ-11B
NAS Lemoore	USN	UCLASS
Pt Mugu	USN	MQ8
San Clemente Island	USA	RQ11B
	USSOCOM	MQ1B
	USSOCOM	MQ9A
SCLA Victorville	ANG	MQ1B
	USA	YMQ-18A
	USSOCOM	A160
	USSOCOM	RQ11B
SSTC, Imperial Beach	USSOCOM	PUMA AE
	USSOCOM	RQ11B
	USSOCOM	Wasp
Simi Valley	USA	RQ11B
	USSOCOM	PUMA AE
	USSOCOM	Wasp
Vandenberg AFB	USSOCOM	RQ11B
COLORADO		
Fort Carson	USA	RQ11B
	USA	RQ7B
	USSOCOM	MQ18
	USSOCOM	MQ9A
Pinon Canyon	USA	RQ11B
	USSOCOM	RQ7B

Location	**Service**	**UAS**
USAF Academy	USAF	Aerosonde
	USAF	Rascal
	USAF	Alpha 60
FLORIDA		
Appalachicola	USSOCOM	Wasp
	USSOCOM	Raven
	USSOCOM	Puma AE
Archer Field	USAF	Wasp
	USAF	MAV
Avon Park	USSOCOM	MQ1B
	USSOCOM	MQ9A
Homestead	USSOCOM	PUMA AE
	USSOCOM	RQ11B
	USSOCOM	Wasp
Hurlburt Fld	USSOCOM	MQ1B
	USSOCOM	MQ9A
	USSOCOM	PUMA AE
	USSOCOM	RQ11B
	USSOCOM	Wasp
NAS Jacksonville	USN	RQ4 Blk10
NOLF Choctaw	USSOCOM	PUMA AE
	USSOCOM	RQ11B
	USSOCOM	Scan Eagle
	USSOCOM	Wasp
NS Mayport	USN	MQ8
Starke	ARNG	RQ11B
	ARNG	RQ7B
GEORGIA		
Fort Kenning	USA	RQ11B
	USA	RQ7B
	USSOCOM	PUMA AE
	USSOCOM	RQ11B

Table 7. (Continued)

Location	Service	UAS
	USSOCOM	Viking
	USSOCOM	RQ7B
Fort Stewart	USA	MOS
	USA	RQ11B
	USA	RQ7B
	USA	MQ1C
	USSOCOM	MQ1B
	USSOCOM	MQ9A
Hunter AAF	USSOCOM	PUMA AE
	USSOCOM	RQ11B
	USSOCOM	RQ7B
	USSOCOM	Viking
HAWAII		
Kaneohe Bay	USMC	RQ-11B
	USMC	RQ-7B
Wheeler	USA	RQ11B
	USA	RQ78
IDAHO		
Boise	ARNG	RQ11B
	ARNG	RQ7B
ILLINOIS		
Havana Airport	ARNG	RQ7B
Ipava	ARNG	RQ11B
INDIANA		
Camp Atterbury	ARNG	RQ11B
	ARNG	RQ7B
Jefferson Range	USSOCOM	PUMA AE
	U5SOCOM	RQ11B
	USSOCOM	RQ7B
	USSOCOM	Viking

Location	**Service**	**UAS**
KANSAS		
Fort Riley	USA	RQ11B
	USA	RQ7B
	USA	MQ1C
KENTUCKY		
Corbin	USSOCOM	PUMA AE
	USSOCOM	RQ11B
	USSOCOM	Wasp
Fort Campbell	USA	RQ11B
	USA	RQ7B
	USA	MQ1C
	USSOCOM	PUMA AE
	USSOCOM	RQ11B
	USSOCOM	RQ7B
	USSOCOM	Viking
Fort Knox	USA	RQ11B
	USA	RQ7B
Lexington	USSOCOM	PUMA AE
	USSOCOM	RQ11B
	USSOCOM	Wasp
LOUISIANA		
Camp Claiborne	ARNG	RQ7B
Fort Polk	ANG	MQ1B
	USA	RQ11B
	USA	RQ7B
	USSOCOM	MQ1B
	USSOCOM	MQ9A
MARYLAND		
NAS Patuxent	USN	RQ4 13(k10
	USN	RQ4A BAMS
	USN	MQ4C BAMS
	USN	MQ8
	USN	UCAS-D
	USN	UCLASS
	USMC	RQ-7B
Webster Outlying Field	USN	MQ8
	ARNG	RQ7B
MASSACHUSETTES		

Table 7. (Continued)

Location	Service	UAS
Camp Edwards	ARNG	RQ11B
MICHIGAN		
Camp Grayling	ARNG	RQ11B
	ARNG	RQ7B
Lake Margretha	ARNG	RQ11B
MINNESOTA		
Camp Ripley	ARNG	RQ11B
	ARNG	RQ7B _.
MISSISSIPPI		
Camp Shelby	ARNG	RQ11B
	ARNG	RQ7B
Stennis Space Center	USSOCOM	PUMA AE
	USSOCOM	RQ11B
	USSOCOM	Wasp
MISSOURI		
Fort Leonardwood	USA	RQ11B
NEVADA		
Creech AFB	USAF	MQ1B
	USAF	MQ9A
	USSOCOM	MQ9A
Desert Rock Airstrip	USSOCOM	Puma
NEW JERSEY		
Warren Grove	ARNG	RQ7B
NEW MEXICO		
Albuquerque	USA	Wasp
Cannon AFB	USSOCOM	MQ1A
	USSOCOM	MQ9A
	USSOCOM	PUMA AE
	USSOCOM	RQ11B
	USSOCOM	Wasp
Holloman AFB	USAF	MQ1B
	USAF	MQ9A
Playas	USSOCOM	PUMA AE
	USSOCOM	RQ11B
	USSOCOM	Wasp
	USA	RQ11B
White Sands	USSOCOM	MQ1B
	USSOCOM	MQ9A
	USSOCOM	RQ7B
NEW YORK		

Location	**Service**	**UAS**
Fort Drum	ANG	MQ9A
	USA	RQ11B
	USA	RQ7B
Seneca	ARNG	RQ11B
NORTH CAROLINA		
MCB Camp Lejune	USMC	RQ11B
	USMC	RQ21A
	USSOCOM	RQ11B
	USSOCOM	Wasp
	USMC	RQ21A
Fort Bragg	USA	RQ11B
	USA	RQ7B
	USA	MQ1C
	USSOCOM	MQ1B
	USSOCOM	MQ9A
	USSOCOM	PUMA AE
	USSOCOM	RQ11B
	USSOCOM	RQ7B
	USSOCOM	Viking
	USSOCOM	Wasp
MCALF Bogue	USMC	RQ7B
	USMC	RQ21A
MCAS Cherry Point	USMC	RQ7B
	USMC	RQ21A
NORTH DAKOTA		
Devils Lake	ARNG	RQ11B
Grand Forks AFB	ANG	MQ1B
	USAF	RQ4 Blk40
OHIO		
Lacarne	ARNG	RQ11B
Wright Patterson AFB	USAF	Wasp
	USAF	RQ11B
OKLAHOMA		
Camp Gruber	ARNG	RQ11B
Fort Sill	ARNG	RQ11B
	ARNG	RQ7B
OREGON		
Boardman Range	ARNG	RQ7B
Camp Rilea	USSOCOM	Wasp
Portland	USSOCOM	PUMA AE

Table 7. (Continued)

Location	Service	UAS
	USSOCOM	RQ11B
	USSOCOM	Wasp
Salem	ARNG	RQ11B
PENNSYLVANIA		
Fort Indiantown Gap	ARNG	ROM
	ARNG	RQ7B
SOUTH CAROLINA		
Charleston	USN	WASP III
North Aux Field	USSOCOM	RQ7B
	USSOCOM	Puma AE
TENNESSEE		
Savannah	USSOCOM	Scan Eagle
TEXAS		
Brownsville	ARNG	RQ11B
Fort Bliss	USA	RQ11B
	USA	RQ7B
	USA	MQ1C
Fort Hood	USA	MQ5
	USA	RQ11B
	USA	RQ7B
	USA	MQ1C
	USMC	RQ7B
	USMC	RQ21A
	USSOCOM	PUMA AE
	USSOCOM	RQ11B
	USSOCOM	RQ7B
Fort Worth	ARNG	RQ11B
Gatesville	USSOCOM	Viking
UTAH		
Camp Williams	ARNG	RQ11B
	ARNG	RQ7B
Dugway	USA	MQ1B
	USA	MQ5
	USA	RQ7B
VIRGINA		
Fort Pickett	USSOCOM	Viking
	ARNG	RQ11B
	ARNG	RQ7B
Fort Story	USSOCOM	PUMA AE
	USSOCOM	RQ11B
	USSOCOM	Scan Eagle
	USSOCOM	Wasp
Fort AP Hill	ARNG	RQ11B
	ARNG	RQ7B
Fort Eustis	USA	RQ11B

Location	Service	UAS
MCB Quantico	USMC	RQ-11B
	USMC	Wasp
	USMC	RQ21A
NALF Fentress	USSOCOM	PUMA AE
	USSOCOM	RQ11B
	USSOCOM	Wasp
	USSOCOM	5100
WASHINGTON		
Fort Lewis	USA	RQ11B
	USA	RQ7B
McChord AFB	USSOCOM	PUMA AE
	USSOCOM	RQ11B
	USSOCOM	Wasp
Yakima Training	ARNG	RQ7B
	USSOCOM	MQ1B
	USSOCOM	MQ9A
WISCONSIN		
Fort McCoy	ARNG	RQ11B
	ARNG	RQ7B
Camp Douglas	ANG	RQ7B
WYOMING		
Camp Guernsey	ARNG	RQ11B
GUAM		
Anderson AFB	USAF	RQ4
	USMC	RQ7B
	USMC	RQ21A
PUERTO RICO		
Salinas	ARNG	RQ11B
OCEANIC		
Worldwide Oceanic	USN	MOB

* Only block 20/40 RQ-4B aircraft remain at Beale AFB after FY2013 budget is authorized (Block 30s cancelled).

SECTION 4. PROGRESS MADE IN PROVIDING THE NUMBER OFFACILITIES NEEDED FOR UAS INVENTORIES TO SUPPORT OPERATIONS AND TRAINING AND THE FUNDING REQUIRED FOR ANY ADDITIONAL FACILITIES

Past, current, and future MILCON projects necessary to support UAS operations and training are presented by each of the Military Departments below:

Air Force

The Air Force UAS MILCON chart (Table 8: Air Force UAS MILCON Projects) shows past, current, and future MILCON projects supporting MQ-1, MQ-9, and RQ-4 UAS funded by the FY 2012 President's Budget.

Additional Air Force Special Operations Command (AFSOC) MILCON includes: a $30.4-million SOF RPA Hangar/AMU Facility at Cannon Air Force Base to replace Hangar 119 and $1.95 million to beddown the 2"d Special Operations Squadron (SOS) once a final location is determined.

Table 8. Air Force UAS MILCON Projects

FY	MAJCOM	Location	Title	Program Amount ($K)	Status
2003	ACC	BEALE	GLOBAL HAWK SQUADRON OPERATIONS/MAINTENANCE FACILITY	$3,670	HIS[1]
2003	ACC	BEALE	GLOBAL HAWK UPGRADE MAINTENANCE DOCK	$4,600	HIS[1]
2003	ACC	BEALE	GLOBAL HAWK DINING FACILITY	$3,470	HIS[1]
2004	ACC	CREECH	PREDATOR SQUADRON OPERATIONS/AM U/HANGAR	$25,731	HIS[1]
2004	ACC	BEALE	GLOBAL HAWK UPGRADE DOCKS	$8,958	HIS'[1]
2004	ACC	BEALE	GLOBAL HAWK DORMITORY (144 RM)	$14,609	HIS[1]
2005	ACC	CREECH	PREDATOR MAINTENANCE COMPLEX	$27,108	CMP[2]
2005	ACC	BEALE	GLOBAL HAWK UPGRADE DOCK 2	$8,320	HIS[1]
2005	ACC	BEALE	GLOBAL HAWK ADDITION TO AGE FACILITY	$1,866	HIS[1]
2006	ACC	CREECH	PREDATOR OPERATIONS FACILITIES	$23,081	CMP[2]
2006	ACC	CREECH	PREDATOR MAINTENANCE AND LOGISTICS COMPLEX	$19,067	CMP[2]
2006	ACC	CREECH	PREDATOR MUNITIONS COMPLEX	$9,237	CMP[2]
2006	ACC	CREECH	PREDATOR TRAINING FACILITIES	$8,732	CMP[2]
2006	ACC	BEALE	GLOBAL HAWK TWO BAY MAINTENANCE HANGAR	$14,058	HIS'
2007	ACC	CREECH	PREDATOR VARIOUS FACILITIES	$26,000	CNS[3]

FY	MAJCOM	Location	Title	Program Amount ($K)	Status
2007	ACC	CREECH	PREDATOR VARIOUS FACILITIES	$23,923	CMP[2]
2007	PACAF	ANDERSEN	GLOBAL HAWK AIRCRAF MAINTENANCE AND OPERATIONS COMPLEX	$52,800	CMP[2]
2007	ANG	MARCH, CA	PREDATOR OPERATIONS & TRAINING COMPLEX	$6,000	CMP[2]
2007	ANG	HECTOR, ND	PREDATOR OPERATIONS COMPLEX	$5,500	CMP[2]
2007	ANG	ELLINGTON, TX	PREDATOR OPERATIONS COMPLEX	$6,000	CMP[2]
2008	ACC	GRAND FORKS	BRAC - CONVERT HANGAR FOR UAV CORROSION CONTROL	$1,280	HIS[1]
2009	ACC	CREECH	UAS OPS FACILITY	$16,145	CNS[3]
2009	ACC	CREECH	UAS DINING HALL	$7,579	CMP[2]
2009	ACC	CREECH	UAS FLIGHT SIM & ACADEMICS FACILITY	$9,127	CNS[3]
2009	ACC	CREECH	UAS 432 WING HQ MISSION SPT FACILITY	$7,000	CMP[2]
2009	ACC	CREECH	UAS MAIN GATE/SEWER TRANSFER STATION INFRASTRUCTURE	$6,500	CNS[3]
2009	ANG	HANCOCK, NY	TFI-REAPER IOC/FOC	$5,000	CMP[2]
2010	ACC	NAS SIGONELLA	GLOBAL HAWK AIRCRAFT MALNT AND OPS COMPLEX	$31,300	CNS[3]
2010	ACC	HOLLOMAN	UAS FTU COMPLEX	$37,500	CNS[3]
2010	ANG ANG	DAVIS MONTHAN, AZ	TFI-PREDATOR BEDDOWN - FOC	$5,600	CNS[3]
2010	ANG	S. CALIF LOG APT,CA	TFI-PREDATOR LRE BEDDOWN	$8,400	CNS[3]
2010	ANG	FT DRUM, NY	TFI-REAPER LRE BEDDOWN	$2,700	DSG[4]
2011	AFSOC	CANNON	UAS SQUADRON OPS FACILITY	$20,000	DSG[4]
2011	ACC	HOLLOMAN	UAS ADD/ALTER MAINTENANCE HANGAR	$15,470	DSG[4]
2011	ACC	HOLLOMAN	UAS MAINTENANCE HANGAR	$22,500	DSG[4]
2011	ACC	CREECH	UAS AIRFIELD FIRE/CRASH RESCUE STATION	$11,710	RTA[1]
2011	USAFE	RAMSTEIN	UAS SATCOM RELAY PADS AND FACILITY	$10,800	DSG[4]
2011	ANG	DAVIS-MONTHAN, AZ	TFI - PREDATOR FOC - INCREASED ORBITS	$4,650	DSG[4]

Table 8. (Continued)

FY	MAJCOM	Location	Title	Program Amount ($K)	Status
2011	ANG	FORT HUACHUCA, AZ	TFI-PREDATOR LRE BEDDOWN	$11,000	DSG[4]
2011	ANG	FT DRUM, NY	TFI - REAPER INFRASTRUCTURE	$2,500	DSG[4]
2011	ANG	ELLINGTON, TX	TFI-ALTER UAV HANGAR	$7,000	DSG[4]
2012	ACC	SIGONELLA NAVAL AS	UAS SATCOM RELAY PADS AND FACILITY	$15,000	DSG[4]
2012	ANG	SPRINGFIELD, OH	ALTER PREDATOR OPERATIONS CENTER	$6,700	DSG[4]
2013	ACC	UNSPECIFIED	MQ-9 PLANNING AND DESIGN	$314	
2013	ACC	UNSPECIFIED	MQ-9 REAPER FACILITIES	$47,750	
2014	ANG	TBD	PREDATOR OPERATIONS CENTER	$10,200	

Table Notes:
1. Historical (HIS)
2. Complete (CMP)
3. Construction (CNS)
4. Design (DSG)
5. Ready To Advertise (RTA)
(Contracting Package is Ready for Bidding/Solicitation)

Army

The Army UAS MILCON listed below shows current and future MILCON projects supporting MQ-1C, MQ-5B, RQ-7B and RQ-11B that are funded through the FY 2012 President's Budget.

MQ-1C Gray Eagle

The Gray Eagle system will be stationed at existing Army Airfields. Table 9: MQ-1C Gray Eagle MILCON) describes the budget for 14 of the 17 Companies (last three Companies are outside the current MILCON funding window).

Each of the identified hangars will house up to three Companies of Gray Eagles.

Table 9. MQ-IC Gray Eagle MILCON

Location	Building	Budget
Fort Huachuca (Schoolhouse)	I Hangar	FY 20 I: $ 10 million
Fort Hood	2 Hangar	FY 2011: $55 million; FY 201 2: $45 million
F011 Riley	I Hangar	FY 201 2: $68 million
Fort Stewart	I Hangar	FY 2011: $47 million; FY 2013 : $20 millio
Fort Bragg	I Hangar	FY 2012: $72 million
Fort Campbell	I Hangar	FY 20 12: $68 millio

MQ-5B Hunter: No new facilities are planned as all three Companies and the training units are already fielded.

RQ-7B Shadow: There are no unit-specific facilities planned for the Shadow Platoon as the system is a subordinate unit whose maintenance and storage facilities are part of their parent company's Tactical Equipment Maintenance Facility. To improve training and reduce maintenance, the Army will prepare field sites in local training areas that are dedicated for the Shadow UAS.

The field site facility includes a 1000-foot-by-50-foot paved landing strip and adjacent support building. The building is a rudimentary structure (3200 square feet) to support Shadow sustainrnent, provide shelter from adverse weather, and secure the platforms.

The concept is for units to occupy the facility on a temporary basis and schedule it as they would a range or training area. It enables launch and recovery under the veil of the installation's restricted airspace and greatly reduces system damage risks during training. Landing strips are funded in FY 2012 for Fort Bragg, Fort Drum, Fort Bliss, Fort Carson, Fort Lewis, Yakima Training Center, Fort Riley; and, in FY 2013, for AP Hill, Atterbury, Fort Chaffee, Fort Dix, Fort Indian Gap, Knox, Korea, Fort McCoy, Orchard Training Area, Fort Richard, Fort Pickett, Camp Ripley, Camp Roberts, and Camp Shelby.

RQ-11B Raven: Due to its small size, dedicated UAS facilities are not required for this system.

Navy

RQ-4A Broad Area Maritime Surveillance Demonstrator (BAMS D): BAMS D aircraft are launched and recovered from a Forward Operating Base (FOB) in the U. S. Central Command (USCENTCOM) and return to NAS Patuxent River, MD, for periodic maintenance. Facilities for BAMS D are temporary in nature (occupying available hangar space at Patuxent River). Following the declaration of initial operating capability of RAMS, the BAMS D UAS are planned to be returned to the tenant command (NAS Patuxent River) for retirement.

MQ-4C BAMS: Table 10: Navy UAS MILCON Projects) shows current and future MILCON projects that are funded by the FY 2012 President's Budget. Additional Navy MILCON is planned outside the FYDP to support continued establishment of UAS capability.

When production commences in FY 2013, BAMS aircraft will initially be located at NAS Patuxent River, MD, for testing. The first two orbits in USCENTCOM and U. S. Pacific Command (USPACOM)will be established with aircraft located OCONUS in FY 2016 and FY 2017. BAMS training and maintenance facilities are currently planned at Beale AFB, CA, and at Main Operating Bases (MOB) NAS Jacksonville, FL, and NAS Whidbey Island, WA. Additionally, FOBs are planned for Andersen AFB, Guam, and NAS Sigonella, Sicily, and at a location in USCENTCOM.

A BAMS UAS test facility is currently under construction at NAS Patuxent River with a completion date of October 2012. Designs are complete for the BAMS Mission Systems Operator training facility at NAS Jacksonville, and construction will begin upon receipt of FY 2012 funding. Three additional projects — USCENTCOM FOB site, BAMS Mission Control Systems Facility at NAS Jacksonville MOB, and RAMS Maintenance Training Facility at Beale AFB — are in the contracting and development process for FY 2013 funding. BAMS' initial USCENTCOM FOB site location is pending host nation notification.

MQ-8B VTUAV-LCS: VTUAV will leverage existing MH-60 support infrastructure as it will also be supporting LCS requirements. Since LCS composite AvDets will normally have the UA only when supporting an LCS, squadrons will not require ramp or hangar space in support of the MQ-8B.

MQ-8B VTUAV — SOF ISR: HQU-10, located onboard NAS North Island, CA, has been identified as the fleet replacement squadron for VTUAV training. No new construction is envisioned to meet VTUAV training requirements.

Unmanned Carrier-Launched Surveillance and Strike (UCLASS): UCLASS air vehicle basing and testing locations have not been selected at this time since the system is pre-milestone A.

Table 10. Navy UAS MILCON Projects

FY	MAJCOM	Location	Title	Program Amount ($K)	Status
2011	NAVAIR	PATUXENT RIVER	BAMS TEST HANGAR FACILITY	$33,304	CNS'
2012	FFC	JACKSONVILLE	BAMS MISSION SYSTEMS OPERATOR TRAINING FACILITY	$4.482	DSG2
2013	FFC	JACKSONVILLE	BAMS MAIN OPERATING BASE MISSION CONTROL SYSTEMS FACILITY	$24,660	
2013	USCENTCOM	VARLOC MIDDLE EAST	BAMS FORWARD OPERATING BASE FACILITIES	$35,900	
2013	FFC	BEALE	BAMS MAINTENANCE TRAINING FACILITY	$17,370	
2014	USPACOM	GUAM	BAMS FORWARD OPERATING BASE	$76,139	
2015	FFC	WHIDBEY ISLAND	RAMS MAIN OPERATING BASE MISSION CONTROL SYSTEMS FACILITY	$28,130	
2015	USEUCOM	SICILY	BAMS FORWARD OPERATING BASE FACILITIES	$29,730	
2015	FFC	BEALE	BAMS MAINTENANCE HUB HANGAR FACILITIES	$50,983	
2016	FFC	BEALE	BAMS FORWARD OPERATING BASE FACILITIES	$35,224	

Table Notes:
1. Construction (CNS)
2. Design (DSG)

USMC

RQ-7B Shadow and RQ-21A STUAS: VMU-1 and VMU-3 are currently based at Marine Corps Base Twenty-Nine Palms, CA. All active-duty VMU squadrons are currently housed in existing facilities that have been modified to support VMU squadron activity. MILCON construction contracts to support the VMU-3 move to Hawaii and the planned VMU-4 move to Camp Pendleton are either in place or pending.

SECTION 5. THE AVAILABILITY OF AIRSPACE, RANGES, AND OTHER INFRASTRUCTURE AT EACH PLANNED UAS LOCATION, AND A DESCRIPTION OF THE STEPS THAT THE SERVICES PLAN TO TAKE TO OVERCOME ANY LIMITATIONS THAT ADVERSELY IMPACT UAS TRAINING

Over the past several years, UAS have become a transformational force multiplier for DoD. The numbers and roles of UAS have expanded dramatically to meet mission demands, and operational commanders have come to rely upon robust and persistent ISR support from unmanned platforms executing their core missions against hostile forces. DoD UAS require routine NAS access in order to execute operational, training, and support missions and to support broader military and civil demands. UAS will not achieve their full potential military utility unless they can go where manned aircraft go with the same freedom of navigation, responsiveness, and flexibility.

As theater forces return and the Military Departments' UAS fleets expand, DoD will require comprehensive continuation and Joint force training in the peacetime environment. Failure to prepare for this eventuality will result in a loss of combat gained experience. As UAS have matured and acquisition programs of record have emerged in all Military Departments, a concerted effort has been made to ensure, wherever practical and possible, that the Departments share logistics costs and burdens to include training and training systems. Below, each Military Department describes planned UAS basing locations and any mitigation plans for adversely impacted UAS training.

U. S. Air Force (USAF)

Each CONUS location that has plans to base MQ-1, MQ-9, or RQ-4 aircraft is listed below with an associated assessment of range and airspace availability.

Beale AFB (RQ-4): Beale AFB currently operates the CONUS-based AF fleet of RQ-4s. Operation from Beale is conducted under a COA issued by the FAA allowing the aircraft to climb into Class A airspace above Flight Level (FL) 180 and transit to operational locations.

Creech AFB (MQ-1 & MQ-9): Creech AFB is located under the restricted airspace of the Nevada Test and Training Range (NTTR), one of the Air Force's largest and most capable range complexes. Other operations from

Creech are conducted under FAA-issued COAs, primarily for MQ-1/9 flights within the Creech AFB traffic pattern in Class D airspace, and for transits to restricted airspace in CA (R-2508) and the Utah Test and Training Range.

Holloman AFB (MQ-1 & MQ-9): An FAA COA is required to utilize runways at Holloman AFB that are not within restricted airspace. A COA is also required to allow MQ-1/9 access to Restricted Area R-5103 B/C southeast of Holloman AFB in order to transit airspace that is not restricted.

Cannon AFB (MQ-1 & MQ-9): RPAs transiting from Cannon AFB to their training location at the Melrose Range operate under an FAA COA requiring ground observers to follow the aircraft when not in restricted airspace.

March ARB/SoCal Logistics Airport (MQ-1 & MQ-9): An FAA COA requires chase aircraft to escort RPAs from Southern California Logistics Airport (formerly George AFB) to the R-2508 Complex associated with Edwards AFB, NAS China Lake, and Fort Irwin.

Fort Drum (MQ-9): The MQ-9 RPAs associated with the Syracuse Air National Guard (ANG) will be based at Fort Drum, NY. Fort Drum is located under the restricted airspace of the Adirondack Range Complex. An FAA COA is required to allow MQ-9 Reaper launch and recovery at Wheeler Sack Army Airfield in order to transit from military Class D airspace to and from the Misty Airspace Complex and the Adirondack Air Traffic Control Assigned Airspace (ATCAA).

Grand Forks (RQ-4, MQ-1 & MQ-9): Chase aircraft or visual observers are required by FAA to mitigate the RPA/UAS lack of a see/sense and avoid capability. In the case of Southern California Logistics Airport (formerly George AFB), no restricted airspace is available for RPAs to operate or transit to R-2508; while at Grand Forks, an airspace proposal has been submitted to establish restricted airspace to support UAS operations west of Grand Forks AFB. RPA operations beyond the Grand Forks AFB traffic pattern are limited until the airspace proposal is approved and charted by the FAA and a supporting COA is developed.

Remote Split Operations (RSO) only: MQ-1 & MQ-9 operations at Ellsworth AFB, SD, and Whiteman AFB, MO, will not have aircraft assigned to their location.

The USAF will require additional airspace access for UAS operations. The current NAS access does not support developmental Sense and Avoid objectives, nor will it support projected training requirements. Without improved NAS access and improved access to special use airspace (SUA), the capabilities of the USAF UAS force will stagnate or degrade, reducing the USAF overall mission effectiveness.

As the UAS force expands and resets from overseas deployments, the demand for airspace or airspace access will increase. Defense Base Realignment and Closure (BRAC) Commission actions and force restructuring have presented an immediate need for suitable testing and training airspace. While the USAF maintains that exclusionary SUA must remain an interim solution, it is not the preferred option due to the lengthy rulemaking process. Consequently, the USAF adamantly supports exploration of all less exclusionary alternatives (e.g., special security instruction airspace, alert areas, terminal radar service areas, Mode C veil over a military operations areas, Federal Aviation Regulation Part 93: Special Air Traffic Rules and Airport Traffic Patterns) that alone or combined with current technology, provide a means to support near-term AF test and training airspace requirements. These interim airspace measures provide immediate improved NAS access, while USAF and FAA work together towards viable long-term and routine sense-and-avoid solutions without undue burden on other NAS stakeholders.

The USAF will limit requests for additional SUA to that required to support combat readiness and only when less exclusionary airspace options are not available or practical.

The USAF is committed to maintaining the safety of the NAS and minimizing impact on civil users while working towards full UAS NAS integration.

Army

The Army has more than 1,800 UAS that are embedded in maneuver units from Platoon through Corps echelons. Army UAS are found at nearly all Army installations. The larger systems, Hunter and Gray Eagle, operate from Army airfields. Shadow UAS launch and recover predominately from field sites located in the local training areas. The hand-launched Raven requires no prepared location from which to conduct operations. For peacetime training, all of these systems operate primarily inside the confines of a military restricted airspace in support of ground maneuver units.

Army UAS have similar operational challenges as other manned aviation platforms such as adequate airspace to maneuver, realistic range targets and sufficient bed down locations. UAS also have the additional limitation of spectrum availability. Of these challenges, spectrum is the most limiting and requires close coordination between all spectrum users to ensure sufficient numbers of frequency sets are available to conduct training. The Army has continued to modernize existing UAS data links in an effort to become more bandwidth efficient; two examples of this are the Digital Data Link (DDL) for the small UAS and the Tactical Common Data Link (TCDL) for Shadow and larger systems. With both data links, modernization allows for more aircraft to operate in the same amount of bandwidth.

While the Army has significant numbers of UAS deployed to combat operations, their return at the end of hostilities will not have a significant negative impact on training. Similar to other Army assets (manned aviation, artillery, ground maneuver units), these units must prioritize and deconflict their requirements for range resources.

Table 11. Locations Requiring COAs

MQ-1C		RQ-7B SHADOW	
Fort Huachuca	AZ	Fort Greeley (Allen AAF)	AK
El Mirage/Grey Butte	CA	Fort Richardson (Bryant AAF)	AK
MQ-5 HUNTER		Fort Wainwright (Husky DZ)	AK
Cochise College	AZ	Redstone Arsenal	AL
Fort Stewart	GA	Whetstone	AZ
Fort Hood	TX	Camp Roberts	CA
RQ-11B Raven (Class G Notification)		Fort Stewart	GA
Fort Wainwright	AK	Wheeler AAF	HI
Simi Valley	CA	Havana	IL
Pifion Canyon	CO	Fort Knox	KY
Ipava	IL	Fort Polk	LA
RQ-11B Raven (Continued)		RQ-7B Shadow (Continued)	
Ft Polk	LA	Camp Grayling	MI
Seneca	NY	Camp Ripley	MN
Camp Gruber	OK	Camp Shelby	MS
Salem	OR	Fort Drum	NY
Brownsville	TX	Ft Sill	OK
Camp Bowie	TX	Fort Indiantown Gap	PA
Camp Swift	TX	Ft Bliss	TX
Yakima Training Center	WA	Ft Hood	TX
		Fort A.P. Hill	VA

Table 11 Locations Requiring COAs) lists the locations where the Army currently conducts operations outside of Restricted Areas that require a COA from the FAA. In the majority of these locations, the purpose of the COA is to transition from the launch site to adjacent Restricted Areas. Additionally, the Raven can be operated using DoD-FAA agreed-to Class G airspace notification procedures for flights flown over Government-owned or -leased land.

Additional efforts to accommodate home station UAS training beyond the Shadow down-range facilities include the following:

1) Targetry. Conventional training range targetry cannot adequately support UAS acquisition and engagement tasks as it is one-dimensional. To provide requisite targetry, the Army has provided targets that are either physical or digital, full-scale representations of the threat to installations. These targets also come with Digital Aviation Gunnery Ranges and Aviation Add-On Packages for existing digital ranges (Riley, Yakima, and Carson Digital Multipurpose Range Complexes). Three-dimensional targets are necessary to support target acquisition, designation, and engagement with captive training missiles.
2) Scoring. Crew tasks must be scored to ensure the laser is on the target and accurate in order to ensure effective Hellfire engagements. While live missiles will not be fired from UAS platforms at home station, laser accuracy must be assessed to ensure precision targeting capabilities.
3) Urban Terrain. Mission challenges in the contemporary environment dictate the availability of urban terrain on home station ranges. To provide these venues, the Army has initiated and successfully enabled fielding of Air-Ground Integration (A-GI) villages. These villages provide 13 modular buildings and urban ballast on live fire ranges. A-GI villages are on the ground at Fort Drum, Fort Bragg, Fort Stewart, Fort Riley, Fort Hood, Fort Bliss, Pohakulua Training Area (HI) and Donnelly Training Area (AK). Further, villages will be fielded to the Yakima and Carson Digital Multi Purpose Range Complex (DMPRC). The Digital Air Ground Integration Range (DAGIR) includes A-GI villages. Units are currently using these villages for manned aerial platform crew qualification and collective A-GI training.

4) Airspace. Thorough analysis has been conducted to evaluate available restricted airspace at installations that support UAS platforms. The challenges can be broken down into two basic areas.
   a) Adequate area to maneuver/train for each installation. This is the length, width, altitudes, and availability of restricted area airspace at the installation.
   b) Adequate/established corridors from the installation's Army Airfield to its restricted airspace. While Raven and Shadow can be supported down range, Gray Eagle requirements dictate launch and recovery from Army Airfields with sufficient runways.

Navy

Navy UAS operator training for Fire Scout, BAMS and UCLASS is planned to be 100-percent simulation based. Fire Scout and BAMS aircraft will generally not be used to meet or maintain training and readiness except in early development before simulators are delivered. Between FY 2012 and FY 2014, Fire Scout training will involve flights at Webster Field. During underway training periods on board Navy ships, Fire Scout will operate in overland and overwater airspace appropriately cleared for UAS operations. Additionally, ship and Strike Group pre-deployment training will require UAS flights in Fleet training areas. Navy is working within the construct of the DoD UAS Airspace Integration Plan to ensure adequate airspace is available for this integrated training.

BAMS: Training for BAMS operators will be done via simulation, utilizing the same computer-based ground stations used to control BAMS. Use of high-fidelity simulation will limit the need to operate the aircraft to only ship and Strike Group integration training.

Regardless of operating location, CONUS or OCONUS, the MQ-4C Mission Control Station (MCS), along with its embedded Mission System Trainer (MST) only requires electrical power and basic infrastructure to provide full functionality for operations and training. Due to its integrated design, the MST does not require an aircraft (or any related airspace, ranges, or flight-related logistic support) in order to provide fully representative operator training.

The Navy is actively engaged with the FAA to obtain COAs for future CONUS sites for BAMS. The current East Coast COA utilized for BAMS-D based from NAS Patuxent River will serve as a model for obtaining authorizations for other CONUS basing locations. With the final approval of CONUS basing sites by Fleet Forces Command, the program office will work with regional commanders and the FAA to obtain COAs to transit through the NAS.

UCLASS: The Navy is actively developing plans for UCLASS training and basing. Due to the immaturity of the program, final decisions have not yet been made.

Marine Corps

All USMC RQ-7B UAS operators (AVOs), maintainers, and unmanned aircraft commanders (UACs) are trained at the U. S. Army's Fort Huachuca, AZ, UAS Training Center under an Interservice Training Agreement in place since 2007.

Marine Corps RQ-7B initial AVO training is nearly identical to the U. S. Army's AVO curriculum and utilizes a combination of classroom instruction and RQ-7B simulator activities at Ft Huachuca. This initial instruction qualifies AVOs to the Joint Basic Unmanned Qualification (BUQ) Level 2[2].

USMC is reviewing a plan to send students for initial UAC training (for non-winged aviators) through the U. S. Air Force's RPA Pilot's flight and instrument qualification courses at Randolph Air Force Base in San Antonio, TX and Pueblo, CO. The USAF RPA Pilot's curriculum uses a combination of classroom instruction, flight simulation, and actual manned aircraft flight to train future UACs. This training will qualify the UACs up to BUQ Level 4.

Almost half of the USMC's RQ-7B Shadow's upper level training syllabus and most refresher training can be accomplished using RQ-7B simulators located at the VMU squadrons. The remainder of the training is conducted using the actual systems operating in support of USMC training events held within the FAA restricted areas that make up the MCB 29 Palms, CA, range complex, MCB Pendleton, the Yuma, AZ range complex, the NAS Fallon, NV range complex, and the MCB Camp Lejeune range complex.

The RQ-21A training syllabus and pipeline are currently under development by NAVAIR PMA 263/205 and the USMC's Training and Education Command (TECOM). It is expected that initially, RQ-21A AVOs and UACs will be drawn from the ranks of qualified RQ-7B operators.

Active FAA COAs allow the transit of VMU-2 aircraft through uncontrolled airspace from the MCAS Cherry Point class "D" surface area to the Restricted Area R5306C/D, and from the NAS Fallon class "D" surface area to the surrounding restricted areas. A ground based sense and avoid (GBSAA) solution is currently being evaluated by the FAA to support the 6 nm transit at Cherry Point. However, a network of 14 ground-based observers is still required to comply with the COA while the evaluation is ongoing.

USMC is actively engaged in its support to the overall DoD-coordinated efforts to shorten and simplify the FAA COA process in order to allow greater unmanned access to the NAS. To support this goal, USMC is pursuing a combination of technical solutions, such as a reliable GBSAA capability, and increased unmanned operator instrument flight training.

SUMMARY

DoD UAS have become a critical component of military operations. Many DoD UAS now require rapidly expanded access to the NAS and international civil airspace to support operations, training, testing, and broader governmental functions. In order for military aircraft to fly routinely in domestic and international airspace, the aircraft must be certified as airworthy, operated by a qualified pilot/operator in the appropriate class(es) of airspace, and comply with applicable regulatory guidance. DoD exercises sole certification authority for its aircraft and pilots/operators, consistent with authority provided in title 10, U. S. Code.

DoD's UAS NAS access methodology uses an incremental approach to provide DoD UAS critical access via given operations profiles prior to implementing a full dynamic operations solution. DoD's immediate focus is gaining near-term mission-critical access while simultaneously working toward far-term routine NAS access. DoD's airspace integration efforts will have positive affordability effects, such as eliminating the cost to study, analyze and complete a COA. Progress will be accomplished through policy and procedural changes, as well as technology and standards developments described in DoD's UAS Airspace Integration Plan. The end state will be routine NAS access comparable to manned aircraft for all DoD UAS operational, training, and support missions.

Additionally, the Office of the Deputy Assistant Secretary of Defense for Readiness, Directorate for Training Readiness and Strategy, is developing a comprehensive DoD UAS training strategy. The strategy will leverage the skills and expertise of each organization and build on foundational efforts already completed or being studied within the Military Departments. The study will investigate and assess the adequacy of existing and forecast joint,

Military Department, and Combatant Commander UAS plans and programs that identify and describe qualification, continuation, and joint training requirements and CONOPS. The strategy will identify and describe individual, unit, and large force training requirements of all groups of UAS. The result will be a UAS Training Roadmap that guides UAS training shortfall and mitigation analyses, provides UAS training recommendations, and proposes investment considerations for the UAS community. The UAS Training Roadmap will serve as a companion piece to the "Unmanned Systems Integrated Roadmap" to provide guidance for efforts related to delivering UAS capabilities to the Warfighter. Phase one of the study will be complete in early 2012 and will serve to identify critical gaps in policy, guidance, and training concepts of operation.

ACRONYM LIST

AAF	ARMY AIRFIELD
ACC	AIR COMBAT COMMAND
AFB	AIR FORCE BASE
AFR	AIR FORCE RESERVE
AFRC	AIR FORCE RESERVE COMPONENT
AFSC	AIR FORCE SPECIALTY CODE
AFSOC	AIR FORCE SPECIAL OPERATIONS COMMAND
ANG	AIR NATIONAL GUARD
ARNG	ARMY NATIONAL GUARD
ATCAA	ADIRONDACK AIR TRAFFIC CONTROL ASSIGNED AIRSPACE
AVDET	AVIATION DETACHMENT
AVO	AIR VEHICLE OPERATOR
B	BASIC
BAMS	BROAD AREA MARITIME SURVEILLANCE
BAMS-D	BROAD AREA MARITIME SURVEILLANCE-DEMONSTRATOR
BRAC	DEFENSE BASE REALIGNMENT AND CLOSURE
BUQ	JOINT BASIC UNMANNED QUALIFICATION
CAP	COMBAT AIR PATROL
CCDR	COMBATANT COMMANDER
CENTCOM	CENTRAL AREA COMMAND
CLS	CONTRACT LOGISTICS SUPPORT

CMP	COMPLETE
CNS	CONSTRUCTION
COA	CERTIFICATE OF WAIVER OR AUTHORIZATION
COE	CENTER OF EXCELLENCE
CONOPS	CONCEPT OF OPERATIONS
CONUS	CONTINENTAL UNITED STATES
DAGIR	DIGITAL AIR GROUND INTEGRATION RANGE
DDL	DIGITAL DATA LINK
DMPRC	DIGITAL MULTI- PURPOSE RANGE COMPLEX
DOD	DEPARTMENT OF DEFENSE
DSG	DESIGN
FAA	FEDERAL AVIATION ADMINISTRATION
FLD	FIELD
FMS	FOREIGN MILITARY SALES
FOB	FOREIGN OPERATING BASE
FOC	FULL OPERATIONS CAPABILITY
FRTP	FLEET RESPONSE TRAINING PROGRAM
FT	FORT
FTU	FLIGHT TRAINING UNIT
FY	FISCAL YEAR
GBSAA	GROUND BASED SENSE AND AVOID
HIS	HISTORICAL
HQ	HEADQUARTERS
ISR	INTELLIGENCE, SURVEILLANCE, AND RECONNAISSANCE
LCS	LITTORAL COMBAT SHIP
LRE	LAUNCH AND RECOVERY ELEMENT
MCAGCC	MARINE CORPS AIR GROUND COMBAT CENTER
MCALF	MARINE CORPS AUXILLIARY LANDING FIELD
MCAS	MARINE CORPS AIR STATION
MCB	MARINE CORPS BASE
MILCON	MILITARY CONSTRUCTION
MOB	MAIN OPERATING BASE
MOS	MILITARY OCCUPATION SPECIALTIES
MRMUAS	MEDIUM- RANGE MARITIME UNMANNED AERIAL SYSTEM
MST	MISSION SYSTEM TRAINER
MTOE	UNIT MODIFIED TABLE OF ORGANIZATIONAL EQUIPMENT

NALF	NAVY AUXILLIARY LANDING FIELD
NAS	IN GEOGRAPHICAL REFERENCE: NAVAL AIR STATION
NAS	IN AIRSPACE REFERENCE: NATIONAL AIRSPACE
NAWS	NAVAL AIR WEAPONS STATION
NOLF	NAVAL OUTLYING FIELD
NS	NAVAL STATION
NTTR	NEVADA TEST AND TRAINING RANGE
PACAF	PACIFIC AIR FORCE
PACOM	PACIFIC COMMAND
POM	PROGRAM OBJECTIVE MEMORANDUM
PT	POINT
RPA	REMOTELY PILOTED AIRCRAFT
SATCOM	SATELLITE COMMUNICATIONS
SCLA	SOUTHERN CALIFORNIA LOGISTICS AIRFIELD
SO	SENSOR OPERATOR
SOF	SPECIAL OPERATIONS FORCES
SOS	SPECIAL OPERAIONTS SQUADRON
SPT	SUPPORT
SQFT	SQUARE FOOT
SSTC	SILVER STRAND TRAINING COMPLEX
STUAS	SMALL TACTICAL UNMANNED AERIAL SYSTEM
SUA	SPECIAL USE AIRSPACE
T/O	TABLE OF ORGANIZATION
TCDL	TACTICAL COMMON DATA LINK
TECOM	TRAINING AND EDUCATION COMMAND
TSRA	TRAINING SYSTEMS REQUIREMENTS ANALYSIS
UAC	UNMANNED AIRCRAFT COMMANDER
UAS	UNMANNED AIRCRAFT SYSTEMS
UCAS	UNMANNED COMBAT AIR SYSTEM
UCLASS	UNMANNED CARRIER LAUNCHED AIRBORNE SURVEILLANCE AND STRIKE
USAF	UNITED STATES AIR FORCE
UDET	UNMANNED DETACHMENT
URT	UNMANNED RPA TRAINING
USA	UNITED STATES ARMY
USAF	UNITED STATES AIR FORCE
USAFE	UNITED STATES AIR FORCE EUROPE
USMC	UNITED STATES MARINE CORPS

USN	UNITED STATES NAVY
USSOCOM	UNITED STATES SPECIAL OPS COMMAND
VMU	VEHICLE MAINTENANCE UNIT
VTUAV	VERTICAL TACTICAL UNMANNED AERIAL VEHICLE
WSMR	WHITE SAND MISSILE RANGE COMPLEX

End Notes

1 MQ-1/9: 4 aircraft per combat air patrol (CAP) and 10 mission control element (MCE) crews per CAP; RQ-4: 3 aircraft per CAP, 15 pilots for MCE, 5 pilots for launch and recovery element (LRE) and 15 sensor operators (SO) per orbit.

2 CJCS 3255.01, "Joint Unmanned Aircraft Systems Minimum Training Standards, July 17, 2009."

INDEX

E

F

G

H

I

N

O

P

Q

R

S

T

U